Grounding Globalization

Antipode Book Series

General Editor: Noel Castree, Professor of Geography, University of Manchester, UK
Like its parent journal, the Antipode Book Series reflects distinctive new developments in radical geography. It publishes books in a variety of formats – from reference books to works of broad explication to titles that develop and extend the scholarly research base – but the commitment is always the same: to contribute to the praxis of a new and more just society.

Published

Grounding Globalization: Labour in the Age of Insecurity
Edward Webster, Rob Lambert and Andries Bezuidenhout

Privatization: Property and the Remaking of Nature-Society Relations
Edited by Becky Mansfield

Decolonizing Development: Colonial Power and the Maya
Joel Wainwright

Cities of Whiteness
Wendy S. Shaw

Neoliberalization: States, Networks, Peoples
Edited by Kim England and Kevin Ward

The Dirty Work of Neoliberalism: Cleaners in the Global Economy
Edited by Luis L. M. Aguiar and Andrew Herod

David Harvey: A Critical Reader
Edited by Noel Castree and Derek Gregory

Working the Spaces of Neoliberalism: Activism, Professionalisation and Incorporation
Edited by Nina Laurie and Liz Bondi

Threads of Labour: Garment Industry Supply Chains from the Workers' Perspective
Edited by Angela Hale and Jane Wills

Life's Work: Geographies of Social Reproduction
Edited by Katharyne Mitchell, Sallie A. Marston and Cindi Katz

Redundant Masculinities? Employment Change and White Working Class Youth
Linda McDowell

Spaces of Neoliberalism
Edited by Neil Brenner and Nik Theodore

Space, Place and the New Labour Internationalism
Edited by Peter Waterman and Jane Wills

Grounding Globalization

Labour in the Age of Insecurity

Edward Webster, Rob Lambert and
Andries Bezuidenhout

Blackwell
Publishing

BLACKWELL PUBLISHING
350 Main Street, Malden, MA 02148–5020, USA
9600 Garsington Road, Oxford OX4 2DQ, UK
550 Swanston Street, Carlton, Victoria 3053, Australia

First published 2008 by Blackwell Publishing Ltd

1   2008

Library of Congress Cataloging-in-Publication Data

Webster, Eddie.
  Grounding globalization : labour in the age of insecurity / Edward Webster, Rob Lambert and Andries Bezuidenhout.
     p. cm. – (Antipode book series)
  Includes bibliographical references and index.
  ISBN 978-1-4051-2914-5 (pbk. : alk. paper) – ISBN 978-1-4051-2915-2 (hardcover : alk. paper)
1.   Anti-globalization movement. 2.   Globalization. 3.   Neoliberalism. 4.   Labor movement.
I. Lambert, Robert. II. Bezuidenhout, Andries. III. Title.

  JZ1318.W433 2008
  331.88–dc22
                                                              2007049846

A catalogue record for this title is available from the British Library.

Set in 10.5/12.5pt Sabon
by SPi Publisher Services, Pondicherry, India
Printed and bound in Singapore
by Utopia Press Pte Ltd

For further information on
Blackwell Publishing, visit our website at
www.blackwellpublishing.com

Contents

Preface
A Journey of Discovery

As we were finalizing this book in November 2006 in Perth, Australia, South Korean scientists revealed that they had found radioactive particles in the air. This, they said, confirmed that North Korea had indeed conducted nuclear tests. Australian troops remain in an occupied Iraq, which looks more and more like the history of Vietnam repeating itself in the farcical way many of us expected. In the region, the anti-terrorist campaign unleashed by the United States after 9/11 is increasingly being used in countries such as the Philippines to assassinate political opponents. Most striking is the deep anxiety over the need to preserve the environment, and water in particular, against the background of dramatic predictions of the impact of climate change. There were even news reports of central rivers running dry. In South Africa, crime is out of hand, with sociologists who attended the World Congress of Sociology returning to their home countries traumatized, not so much because of crime, but because the security guards at the doors of their hotels did not allow them to venture out to see the beautiful city of Durban on the Indian ocean with its cosmopolitan mix of African, Asian and European cultures. This state of siege and paranoia may be understandable when one considers the country's painful history and the fact that the unemployment rate still hovers around a staggering 40 percent. Many South Africans have immigrated to Perth in Western Australia to escape South Africa's high levels of violent crime. In short, we live in an age of insecurity.

Faced with insecurity, persons tend to retreat into the familiar – their country, their neighbourhood, their home, their family – and sometimes their 'race'. Indeed, at times when the world faced similar levels of insecurity, we saw the rise of some of the worst atrocities of human history. One author who reflected on such times was Karl Polanyi, who wrote his major work at the end of the Second World War. At the forefront of his mind was the rise of fascism. Why do people turn to fascist leaders, and under what conditions does fascism become salient as a political ideology? It is no wonder that people are returning to Polanyi in order to make sense of current times of insecurity.

But a person's sense of security or insecurity does not only relate to threats of terrorism, war and the pending environmental catastrophe. It also has to do with social and economic security: job security, household income, feeding children, improving current living conditions. Insecurity in these facets of existence is not new. Capitalist development produced dispossession and insecurity. The great depression of the 1930s undermined these very facets of human existence for millions across the globe. We will show that what is new is the strategy of neoliberalism to *consciously manufacture* insecurity as a strategy to undermine the collective power of civil society movements. To be sure, the two realms mentioned here are closely linked, as Polanyi showed shortly after the Second World War. This is where this book began; how societies respond to the restructuring of work – the *global* restructuring of work – and what possibilities there are for a democratic outcome. Polanyi called this the double movement; how under the impact of what he called the Great Transformation in nineteenth-century Europe, societies protected themselves by subordinating the market to society.

We argue that the world is in the grip of a Second Great Transformation, a neoliberal project that is the dominant policy paradigm in the national affairs of all three countries that we have chosen to examine. But, just as Polanyi anticipated that the pendulum would swing against unregulated markets, our findings detect a similar shift from market fundamentalism towards the need to protect society and the environment against an unregulated market.

As we were writing this preface the then shadow minister for foreign affairs and now Prime Minister of Australia, Kevin Rudd, wrote that John Howard, Australia's conservative Liberal Prime Minister,

[was] in the process of unleashing new forces of market fundamentalism against youth workers; families trying to spend sufficient time together; and communities trying to negotiate with single, major employers experimenting with their new-found powers. Breadwinners are now at risk of working less predictable shifts, spread over a seven day week, not sensitive

to weekends, and possibly for less take-home pay. The pressures on relationships, parenting and the cost and quality of child care are without precedent ... Howard has never accepted that labour is different to any other commodity: it too is something whose value should be determined on a free market. (Rudd 2006: 49–50)

In what could be the beginnings of a more serious challenge to market fundamentalism in Australia, Kevin Rudd suggests that 'the opportunity arises for Labour to reclaim the centre of Australian politics, thereby reframing the national political debate. Labour also now has the opportunity to form fresh political alliances with other groupings alienated by this new form of market fundamentalism, which is blind and indifferent to its social consequences' (Rudd 2006: 50)

In Korea,[1] at the time of writing this preface, such opportunities were being taken as 138,000 workers participated in a 4-hour 'warning strike' to stop, amongst other demands, the Irregular Workers' Bill designed to further liberalize the labour market. In conjunction with this strike, trade unions from 40 countries participated in solidarity actions as part of an international day of action to support South Korean workers. This was followed a week later by a general strike in Korea.

In South Africa there are significant signs that government is shifting from its narrow focus on global integration and competitiveness to acknowledge that the state will have to redistribute resources actively in an effort to overcome the social crisis caused by poverty. This resulted, in 2004, in the announcement of an expanded public works programme which intended to create one million temporary jobs over a five-year period, a modestly increased budget deficit, and a shift in emphasis from privatization to the state's role in stimulating economic activity.

The World Social Forum, established in 2001 as a counter to the neoliberal World Economic Forum, is the clearest example of the double movement in action. A network of NGOs, social movements, and as yet hesitantly, the labour movement, it presents a global challenge to the domination of multinational and financial capital. Indeed, the 'rediscovery of poverty' by the international financial institutions, the adoption of the Millennium Development Goals by the United Nations and the shift from market-led structural adjustment to Poverty Reduction Strategies could be seen as a classic example of Polanyi's notion of the

---

[1] Korea refers to 'South Korea' throughout the book, unless specified differently.

pendulum swinging on a global scale against market fundamentalism. But, as the ILO argues:

> The economy is becoming increasingly global, while social and political institutions remain largely local, national and regional. None of the existing institutions provide adequate democratic oversight of global markets, or redress basic inequalities between countries. The imbalances point to the need for better institutional frameworks and policies if the promise of globalization is to be realized. (ILO 2004: 3)

Researching and writing this book has been a long journey. We draw on Polanyi to go beyond him, to locate his problematic in the context of globalization. It has been said that the true voyage of discovery is not to travel to foreign places, but to see reality through new eyes. This is indeed true of our journey: our most exciting discoveries have not derived from visiting unfamiliar places, but through gaining new insights into the contemporary world. What are these insights?

The journey began in 2000 when two of the authors were invited to contribute to a study on what was far-sightedly titled *Reinventing social emancipation: studies in counter-hegemonic globalization*. Rob Lambert, who doubles up as a professor at the University of Western Australia with the role of International Affairs Officer of Unions Western Australia, had in 1990 initiated a response to Australia's rapid liberalization that brought together democratic unions in the global South to promote trade union rights. It was called the Southern Initiative on Globalization and Trade Union Rights (SIGTUR) and we, that is, Rob Lambert and Edward Webster, contributed to this project on SIGTUR as an example of the new labour internationalism (Lambert & Webster 2003, 2004, 2006; Webster & Lambert 2004).

Our analysis of the new labour internationalism was to decisively shape the direction of this book. In the first instance it became clear to us that transnational movements will not work without being embedded, or grounded, in local society. Foregrounding *place* drew us into the exciting intellectual work of the new labour geography. We were delighted when we were invited to contribute to a special volume of the radical journal of geography *Antipode* on the new labour internationalism (Lambert & Webster 2001; Waterman & Wills 2001). The publication of this article led to an invitation to contribute a book-length manuscript to the new *Antipode* Book Series, which aims to 'further geographical scholarship that takes issue with the existing order of things'. Our contribution to radical geography is our focus on work and place, and the relationship between corporate power, scale and space; in particular the opportunities opened up by this approach for constructing a global counter-movement. It led us to ground the study in three distinct places, the towns of Ezakheni in South Africa, Orange in Australia, and Changwon in South Korea.

A second impact was the identification of new sources of power. Labour's traditional sources of power are being eroded by the new employment relationships and labour law reforms introduced under the political project of neoliberalism. This led us to the exciting new labour scholarship emerging in the United States and its use of the concept of social movement unionism. It also confirmed our view that the restructuring of the world of work was the lens through which we could best interrogate the Polanyi problem in the age of globalization. Fortunately both Andries Bezuidenhout and Rob Lambert had recently completed in-depth studies of the restructuring of the white goods industries in South Africa and Australia (Bezuidenhout 2004, 2005b; Lambert 2005; Lambert, Gillan & Fitzgerald 2005; Lambert & Gillan 2007). Both studies revealed a common process of restructuring and growing concentration of the global white goods industry. We decided to include South Korea as LG, one of the leading white goods transnational companies globally, is an example of a successful late-industrialiser that has generated strong opposition in recent years in its attempt to create a more flexible labour market.

What emerged from a more careful reading of Polanyi were major gaps in his account of the double movement, in particular his under-theorizing of how a counter-movement is constructed. This led us to our third insight: the absence of a theory of social movements in Polanyi. Fortunately, transnational activism has become a subject of serious study in recent years and we were able to draw on this work, and Sidney Tarrow in particular, to identify the processes through which transnational activism emerges.

A final and quite fundamental insight was the need to rethink labour studies. What has been missing from the study of labour in the global economy is the impact of global restructuring on the non-working life of workers. The reconfiguration of the employment relationship – through the growing casualisation of work, downsizing and retrench-ment – impacts directly on workers' households and the communities of which they are part. To understand labour in the global economy, we need to change the way we study labour. Labour studies should not be an analysis of the workplace only; we need to examine workers as a totality, workers in society. To understand these responses we surveyed workers not only in the workplace, but also their households and communities. Through following workers into their homes and communities, the real differences in the working lives of those in Ezakheni, Orange and Changwon emerged.

To take labour seriously in the era of globalization is, then, to take society seriously. As Burawoy argues, 'the post-communist age calls for a Sociological Marxism that gives pride of place to society alongside, but distinct from, the state and the economy' (Burawoy 2003: 193). For Polanyi, the expansion of the market threatened society, which reacted by reconstituting itself as 'active society', thereby harbouring the embryo of a democratic socialism. But our findings on society – its segmentation, its racism, its criminality and its patriarchy – are at odds with Polanyi's solidaristic notion of society. As Burawoy observes: 'For Polanyi the battle between society and the market is a battle of the Gods, between good and evil' (Burawoy 2003: 248).

It would be remiss if we were not to acknowledge the key role that engaged intellectuals make in the construction of a counter-movement. It was the activist-scholar Richard Turner, banned and later assassinated in 1978 by the Durban security police, who first persuaded us of the importance of utopian thinking and it is to his ideas and life's work that we dedicate this book. We draw on Turner, not out of nostalgia, but as an example of an activist-scholar, who was able to combine utopian thinking with practical social action. Indeed, he was a forerunner of what is celebrated today as public sociology.

For us, the journey is not over; what we identify in this book is a potential counter-movement to the hegemony of neoliberalism. We have provided a theoretical framework for taking this project forward, not a blueprint for action. We focus in this book primarily on the commoditization of labour, not the commoditization of nature and the ecological crisis that this is generating. This most critical of issues must be left to another occasion, but what needs to be emphasized is that, like Polanyi, we identify different responses to the commoditization of social life. Retreat from the market is one; mobilization against it is another. What form such mobilization will take remains for us an open question.

However, what is unequivocal is that the liberal argument that the rising density, since around 1980, of economic integration across national boundaries and the ideological shift towards neoliberalism with its foregrounding of market-led development over social equity, is not creating a global 'level playing field'. The percentage Gross National Product (GNP) per capita as a percentage of the core's GNP per capita income declined between 1960 and 1999 in sub-Saharan Africa, Latin America, North Africa and West Asia (Wade 2003: 26). The only part of the world where incomes have risen is East Asia, where countries such as South Korea went against the policy paradigm of market-led development. The twin processes of intensified economic integration and ideological neoliberalism, what has become known as globalization, has increased social inequality within developed and developing countries alike, creating a heightened sense of insecurity.

In this study we examine three paths from the periphery where the state has facilitated industrialization: Australia, an advanced industrial resource-based country with a growing service economy; Korea, a late industrialiser, based on the export of consumer and capital goods; and South Africa, an enclave economy, founded on a dual logic of inclusion and exclusion, based, as with Australia, on resources with a growing service economy.

The most difficult challenge this book presented is the process of writing itself. This is not a book written by three separate authors; it has been written most of the time by three authors together. Of course at times it has been written separately, and we have communicated with each other in cyberspace, but mostly it has been written face to face by all three of us together. Writing collectively can be emotionally demanding, as authors contest the form and content of the written word, but it undoubtedly added enormous value to the final product. This is because it brought together not only three different intellectual biographies, but the interaction between them, thereby adding a fourth author, the outcome of the interaction itself. It is this creative synthesis between us that has been the most difficult but unquestionably the most rewarding part of the project. Perhaps one of the most influential insights to emerge out of these encounters was a recognition of the distinctiveness of the counter-movement concept as a social phenomenon greater than simply trade unions in search of re-empowerment. We came to envisage the counter-movement as a creative synthesis of different types of social movement, potentially generating a new social force requiring a new form of politics. We concluded that there is no more critical task than the active engagement of intellectuals, within and outside of the labour and other social movements, in defining the form and character of such a movement.

We have accumulated a number of debts on this journey. Our findings and insights on Korea would not have been possible without Yoon Hyowon whom we employed from September to November 2005 to undertake a survey of the LG plant in Changwon. Fortunately, we were able to work with him in Johannesburg in November that year to write up the results of his survey. Yoon also conducted field work on privatization and the anti-privatization struggles in Korea. We would also like to thank Michael Burawoy for helping us clarify the structure of the book at a crucial stage in ordering our analysis. We have also drawn the first part of the title of our book, 'Grounding Globalization', from the concluding chapter of his edited volume *Global Ethnography* (see Burawoy et al. 2000).

Khayaat Fakier and Lindiwe Hlatswayo assisted us in undertaking the survey of households and communities in Ezakheni. Maria van Driel provided us with a valuable report on privatization in South Africa. We would also like to acknowledge the assistance of Janaka Biyanwila for researching privatization in Australia and Mike Gillan on researching restructuring of the white goods industry in Australia. We thank Noel Castree for inviting us to contribute to this series and Hannchen Koornhof, Jack Messenger and Jacqueline Scott who guided the book through to production.

We gratefully acknowledge South Africa's National Research Foundation (NRF), which provided the real financial capacity to realize this ambitious comparative project, as well as the Australian Research Council for its contribution to this project, which enabled empirical work to proceed in South Africa, Australia and Korea, enriched by face-to-face encounters by the authors in South Africa and Australia on three separate occasions.

Abbreviations

ACTU	Australian Council of Trade Unions
ALP	Australian Labour Party
AMWU	Australian Manufacturing Workers' Union
ANC	African National Congress
APF	Anti-Privatisation Forum
AWAs	Australian Workplace Agreements
AWU	Australian Workers Union
BHP	Broken Hill Proprietary
CBOs	Community Based Organizations
CFMEU	Australian Construction, Forestry, Mining and Energy Workers Union
CITU	Centre of Indian Trade Unions
COSATU	Congress of South African Trade Unions
CR	Cost Reduction
DLP	Democratic Labour Party
DTI	Department of Trade and Industry
DWAF	Department of Water Affairs
EME	Edison Mission Energy
ESI	Electricity Supply Industry
ESKOM	Electricity Supply Commission
ETUC	European Trade Union Confederation
FKTU	Federation of Korean Trade Unions
GATS	General Agreement on Trade in Services
GATT	General Agreement on Tariffs and Trade
GDP	Gross Domestic Product
GEAR	Growth, Employment and Redistribution
GLU	Global Labour University

GURN	Global Union Research Network
HDI	Human Development Index
ICEM	International Chemical, Energy and Mining Workers' Federation
ICF	International Federation of Chemical and General Workers Unions
ICFTU	International Confederation of Free Trade Unions
IFP	Inkatha Freedom Party
IIE	Institute of Industrial Education
ILO	International Labour Organization
IMF	International Metalworkers Federation
IMF	International Money Fund
ISCOR	Iron and Steel Corporation
ITS	International Trade Secretariats
ITUC	International Trade Union Confederation
IUF	International Union of Food, Agricultural, Hotel, Restaurant, Catering, Tobacco and Allied Workers' Associations
IWMA	International Working Men's Association
KCTU	Korean Confederation of Trade Unions
KCWC	Korea Contingent Worker Centre
KECO	Korea Electric Company
KEPCO	Korea Electric Power Corporation
KLEA	Korean Labour Education Association
KMI	KwaZulu Marketing Initiative
KPPIU	Korea Power Plant Industry Union
KTUC	Korean Trade Union Congress
LGE	LG Electronics
LGEI	LG Electronics Investment
LPM	Landless People's Movement
MAU	Maritime Union of Australia
MNC	Multinational Corporation
NCKTU	National Centre of Korean Trade Unions
NCP	National Competition Policy
NEDLAC	National Economic Development and Labour Council
NEM	National Electricity Market
NER	National Electricity Regulator
NFA	National Framework Agreement
NGOs	Non-Governmental Organizations
NP	(Afrikaner) National Party
NPCC	Northern Province Council of Churches
NPOs	Non-Profit Organizations

NRF	National Research Foundation
NSW	New South Wales
NUM	National Union of Mineworkers
NUMSA	National Union of Metalworkers of South Africa
OECD	Organisation for Economic Cooperation and Development
PAC	Pan-African Congress
PPPs	Private Public Partnerships
RDP	Reconstruction and Development Program
RED	Regional Electricity Distributors
RGUN	Rio Global Union Network
RSI	Repetitive Strain Injury
SABC	South African Broadcasting Corporation
SAMWU	South African Municipal Workers' Union
SECV	State Electricity Commission of Victoria
SEIU	Service Employees International Union
SEWA	Self Employed Workers Association (India)
SEWU	Self Employed Women's Union (South Africa)
SIGTUR	Southern Initiative on Globalisation and Trade Union Rights
SOEs	State Owned Enterprises
STCs	Short-Term Contractors
TAC	Treatment Action Campaign
TDR	Tear Down and Redesign
TNC	Transnational Corporation (as opposed to MNC when ownership resides mainly in one country, but operations are transnational)
TUAC	Trade Union Advisory Council
UIM	(Protestant) Urban Industrial Mission
UNDP	United Nations Development Programme
WCC	(Electrolux) World Company Council
WCL	World Confederation of Labour
WFTU	World Federation of Trade Unions
WSSA	Water Services of South Africa
WTO	World Trade Organization

1

The Polanyi Problem and the Problem with Polanyi

The Polanyi Problem

In his account of the way in which globalization is 'flattening the world', Thomas Friedman introduces, halfway through his book, a note of caution. He mentions a conversation with his two daughters in which he bluntly advises them: 'Girls, when I was growing up, my parents used to say to me, "Tom, finish your dinner – people in India and China are starving." My advice to you is: Girls, finish your homework – people in China and India are starving for your jobs' (Friedman 2005: 237).

But of course the world – or more specifically, the global economy – is not flat; it is highly uneven. This anecdote captures in a nutshell the widespread insecurity that unevenness creates. Indeed, unevenness and growing insecurity is the central theme of our book. On the one hand the new economy has created unprecedented opportunities for wealth creation, while on the other hand its uneven nature threatens established livelihoods. This implies that people in various parts of the world experience the dislocation brought about by globalization in different ways. Working people in the industrialized North are concerned about their jobs moving to other countries. Major regions of the world economy that were previously insulated from capitalism are now drawn in as major sites of industrial investment. New working classes are created. Other regions of the globe are essentially excluded from these new waves of economic transformation and remain marginal. This is the nature of capitalist industrialization.

Our research across the three nations shows how insecurity is manufactured by market liberalization. Mpumi Khuzwayo is a contract worker at the Defy refrigerator factory in Ezakheni, whose day-to-day routine is riddled with the insecurity of not knowing if her job will be available tomorrow. She saves all the money she can manage for times when she does not work, which is most of the year. She is scared of walking home at the end of her night shift because of high levels of violent crime. She feels that the ruling party, the African National Congress (ANC), has become aloof and removed from its base, but that the populist leader Jacob Zuma can bring the movement back to its roots.

Peter Tyree has worked in the Electrolux plant in Orange for most of the past 25 years of his working life. He was a strong supporter of the Australian Labour Party (ALP) and is an active union member. He has deep roots in Orange and is actively involved in the local football team. Electrolux has retrenched most of his mates and he does not know whether what is left of the factory will remain. His feelings of insecurity led him to support Pauline Hanson's One Nation Party, which did not only advocate stricter immigration laws, but also government protection of local industry.

Bae Hyowon is president of his union, an enterprise union, in Changwon. This company supplies parts to the nearby LG factory. Workers are worried that their company will lose its contract with LG if they openly affiliate to one of the national union federations. LG has increasingly moved its contracts to Chinese suppliers, and Bae Hyowon is experiencing deep feelings of insecurity. He works overtime regularly, including Saturdays and some Sundays. This leaves little free time for his union activities and his involvement in an organization supporting the reunification of North and South Korea. Furthermore, he has little time for his preferred leisure activity, mountain climbing.

Mpumi, Peter and Hyowon experience insecurity as an individual matter, what C. Wright Mills (1959) calls a 'personal trouble'. But when large numbers of workers in different factories, in different countries, experience the same feelings of insecurity, it is no longer a personal trouble only, it is a public issue. To understand insecurity, it is not enough to identify the sense of helplessness, fear, depression, anger, and sadly, often self-destructive behaviour such as suicide, substance abuse and domestic violence. It is necessary, as Mills argued, to identify the broad social forces, institutions and organizations that manufacture this insecurity.

Indeed, the insecurity created by rapid, unregulated social change lies at the centre of the social science project. It had its beginnings in attempts by classical social thinkers, such as Karl Marx, Emile Durkheim and Max Weber, to interpret the 'First Great Transformation' that led to the market economy. Its emergence was connected to the widespread concern with the economic, social, cultural and moral effects of moving from a non-industrial to an industrializing society. This concern reflected the major fault-line of politics at the time between the proponents of economic liberalism and their advocacy of the self-regulating market on one side, and on the other, those who favoured intervention to 'protect society'. The idea of protecting society was not only a radical idea; it was also at the centre of the conservative ideas of Edmund Burke, for example, and his notion of an organic society.

This classical outlook of social science shared the view that work was the fundamental social experience. Work and the social relations structured around work, including the grounding of human life-forms in nature, were seen as the central dynamic of modern industrial society. In recent times, with the turn to the study of culture and consumption and the rise of post-modern obsessions with subjectivity, this concern has tended to fall by the wayside. Also, the labour movement is often relegated to the past and is seen as tired and old. Nevertheless, as the International Labour Organization (ILO)'s Commission on the Social Dimension of Globalization argues: 'Work is central to people's lives. No matter where they live or what they do; women and men see jobs as the "litmus test" for the success or failure of globalization. Work is the source of dignity, stability, peace, and credibility of governments and the economic system' (ILO 2004: 6).

The rapid growth of economic liberalism over the past 25 years has led to the current period of world history being defined as a Second Great Transformation (Munck 2002). The theoretical work of Karl Polanyi is influential in the construction of a sociology of this transformation (Peck 1996; Burawoy et al. 2000: 693; Burawoy 2003; Silver 2003; Munck 2004; Harvey 2006: 113–115). The starting point for an understanding of Polanyi's work is his concept of 'embeddedness' – the idea that the economy is not autonomous, but subordinated to social relations. This is a direct challenge to economic liberalism, which rests on the assumption that the economy automatically adjusts supply and demand through the price mechanism. The idea of a fully self-regulating market economy, Polanyi argued, is a utopian project. In the opening page of Part One of *The Great Transformation* he writes: 'Our thesis is that the idea of a self-adjusting market implied a stark utopia. Such an institution could not exist for any length of time without annihilating the human and natural substance of society' (Polanyi 2001 [1944]: 3–4).

This is the Polanyi problem: creating a fully self-regulated market economy requires that human beings, nature and money be turned into pure commodities. But, he argues, land, labour and money are fictitious commodities, because they are not originally produced to be sold on a market. Labour cannot be reduced simply to a commodity, since it is a human activity. Life itself is not sustained by market forces, but is reproduced socially: in households, in communities, in society. Land is not simply a commodity, because it is part of nature. So, too, is money not simply a commodity, because it symbolically represents the value of goods and services. For this reason, Polanyi concludes, modern economic theory is based on a fiction, an unrealizable utopia.

In his classic study of the industrial revolution Polanyi (2001) showed how society took measures to protect itself against the disruptive impact of unregulated commoditization. As we mentioned in the preface, he conceptualized this as the 'double movement' whereby ever-wider extensions of free market principles generated counter-movements to protect society. Against an economic system that dislocates the very fabric of society, the social counter-movement, he argued, is based on the 'principle of social protection aiming at the conservation of man and nature as well as productive organization, relying on the varying support of those most immediately affected by the deleterious action of the market – primarily but not exclusively, the working and the landed classes – and using protective legislation, restrictive associations, and other instruments of intervention as its methods' (Polanyi 2001: 138–139).

Polanyi's theory is profoundly shaped by moral concern over the psychological, social and ecological destructiveness of unregulated markets. This assessment resonates today because such a relentless drive towards a market orientation lies at the very heart of the contemporary globalization project. As a consequence, market-driven politics dominates nations across the globe (Leys 2001). The discourse of this politics centres on the language of the market: individualism, competitiveness, flexibility, downsizing, outsourcing and casualization.

With reference to the First Great Transformation, Polanyi countered this discourse with a language growing out of a new ethics that challenges the market definition of persons and society. Such a definition reduces all human encounters to relationships between commodities in a conception where, as Margaret Thatcher argued, society does not exist. Polanyi's moral intervention is grounded in the notion of the innate value of persons, hence the centrality of constructing a just and free society, where participatory democracy, at work and in society, recognizes the rights of persons and their communities. In this vision, persons, communities and society are the priority. Thus markets have to be socially regulated. Within such a structuring of social relations, society asserts its control over markets to counter the corrosive effect of insecurity.

Polanyi's work also contains a warning. Insecurity may not necessarily result in progressive counter-movements. It could, and it has, led to its opposite. Indeed, this was the central preoccupation of Polanyi's classic work, namely that the unregulated liberalization of markets between 1918 and 1939 would lead to the rise of fascism. It was this response to liberalization that led to Polanyi's concern with democracy.

Polanyi recognized different responses to the commoditization of labour, but did not explain how these responses come about. The challenge is to identify the responses that are emerging today as Polanyi under-theorized how counter-movements are constructed. In order to understand the Second Great Transformation, and actual and potential counter-movements, we have to address these theoretical shortcomings. We identify five areas of under-theorization in Polanyi's work.

The Problem with Polanyi

The society problem

Polanyi makes constant reference to society, but at no stage is the nature of this central concept clarified. In this, Polanyi is not alone, for as Fred Bloch points out in a comment to Michael Burawoy: '[Your paper] points to the absurdity that the sociological tradition has failed for a hundred and fifty years to give us an adequate or useful conceptualization of "society" – ostensibly the main object of its analysis' (Burawoy 2003: 253). For Burawoy, society occupies a specific institutional space between the state and the economy. He calls this reappropriation of the analysis of society *Sociological Marxism*. Society is not a timeless notion but a specific historical product. Nor is it 'some autonomous realm suspended in a fluid of spontaneous value consensus, rather it is traversed by capillary powers, often bifurcated or segmented into racial or ethnic sectors, and fragmented into gender dominations' (Burawoy 2003: 199). It is also, Burawoy suggests, Janus-faced: 'on the one hand acting to stabilize capitalism but on the other hand providing a terrain for transcending capitalism' (Burawoy 2003: 199).

We would agree with Burawoy when he says that society occupies a certain institutional space between the state and the economy. But how such institutions relate to the state and the economy is contested. Schools may be privatized; hospitals can be run and funded by the state, churches or business; sport can be commercialized. The boundaries between society, the state and markets may be analytically distinct, but in reality these boundaries are not fixed and tend to shift over time.

In order for a society to exist, social relations need to have a certain density of ongoing institutional interaction – a social structure. These structures, Burawoy suggests, tend to include and exclude categories of individuals on the basis of social characteristics and distribute power unequally along similar lines. These characteristics are generally an integration of gender, race, generation and, of course, class inequalities. Social inclusion and exclusion can be constituted on the basis of a certain territory (i.e. spatially), or these boundaries can be drawn *within* a certain territory (i.e. socially). We stress the spatial constitution of society as a key characteristic. Society also constitutes a public domain, where issues are discussed and debated. As in the case of other social institutions, this domain and the rules of engagement within it, are contested.

At the centre of Polanyi's notion of society is that of a contradictory tension with the market: on the one hand markets destroy, undermine, fracture and fragment society, while on the other hand they also create, what he calls *active* society, where individuals come together in groups and movements, generating cultures of solidarity and resistance. An active society is best understood when contrasted with a passive society. We identify two types of passive society: the first is one where the market dominates through the promotion of individualism and consumerism. This reflects a typical neoliberal order, where corporations capture a state in order to search for new areas of profitable investment through the privatization of a range of institutions such as schools and hospitals, and the provision of water and electricity. Corporations have, therefore, both ideological and financial reasons for wanting to control the media and shape the public discourse. In Australia, for example, the media tycoon Rupert Murdoch controls, through News Corporation, 70 per cent of the national newspaper market and, with the new media law amendments, he can extend his control to television and radio stations (Manne 2006: 10).

A second type of passive society is one where the state dominates society. This takes place under authoritarian political regimes, be they of fascist or state socialist nature. The state attempts to control social institutions such as schools, trade unions and other civil society organizations. It is not in the interest of the state to allow open debate in the public domain and hence it tightly controls the media. In fascist states, the ideology operates on the basis of social characteristics and minorities are cast as scapegoats. In such cases society tends to be weak. As Burawoy convincingly shows, faced with the introduction of markets through shock treatment, post-communist Russian society evidenced signs of 'involution', a retreat to the household economy and a barter economy. The result was that Russia could not forge a social response to protect itself against the destructive elements of markets (Burawoy 2001).

A further example of societies where the state dominates is that of colonialism, an area largely unexamined by Polanyi in *The Great Transformation*, whose focus is, with one exception, mainly on the industrialized North. The exception is his deep interest in pre-colonial societies where he identifies non-market relations of exchange as the basis for building alternative social relations based on reciprocity and redistribution. Under colonialism the state is created and captured by the colonizer who proceeds then to impose on the colonized a sharply unequal and racially segmented state. This bifurcated state, in the words of Mamdani, has a contradictory effect: it both destroys indigenous society while at the same time preserving selectively traditional structures that can be manipulated to reproduce cheap labour power (Mamdani 1996). This generates a powerful counter-movement, not against the market as such (indeed, in some cases it may demand an extension of economic opportunities through removing restrictions to trade, produce and to selling labour on an open market), but against colonial rule.[1] This movement, a national liberation movement, aspires to the construction of a new post-colonial society.

For us then, society is not held together by shared values; on the contrary, it reflects an ongoing ideological contest between these different visions of society. What of his alternative? For Polanyi, socialism is the alternative to the self-regulating market. Socialism is essentially, he writes, 'the tendency inherent in an industrial civilization to transcend the self-regulating market by consciously subordinating it to a democratic society' (Polanyi 2001: 242). Although for Polanyi there was no simple teleological transition from capitalism into socialism, it remained unclear how he envisaged such an alternative would emerge.

The spontaneity problem

'Counter-movements cannot be seen as spontaneous, practically automatic responses; they are constructed', argues Munck (2004: 257). There is no consideration in Polanyi's work of how working classes are made and unmade (Silver 2003). In particular, there is no understanding of how the formation of a working class is an active process (Thompson 1963). As Burawoy (2003: 221) observes: 'He [Polanyi] was writing before Edward Thompson's transformative *The Making of the English Working Class*, which underlines the importance of working class traditions for class formation, in particular those of the "free-born Englishman." For a class to mobilize, it needs "resources" – cultural, political and economic. It needs capacity. In Polanyi's account, where might such resources come from?' He concludes, 'the English working class could not be regarded as a blank slate, defenceless against market forces. It was already embedded in community, which gave it the weapons to defend itself and advance active society in its own name' (Burawoy 2003: 222).

To explain the emergence of counter-movements one needs a theory of social movements. Social movement theory provides us with an understanding of the structural conditions, political opportunities and repertoires that movements draw on, and how resources are mobilized when social movements engage in contentious politics (Tarrow 1994; Jenkins & Klandermans 1995; Tilly 2004). Transnational social movements are not merely a reflex against globalization. They are shaped by changes in the opportunity structures of international politics. 'If globalization consists of increased flows of trade, finance, and people across borders', Tarrow argues, 'internationalism provides an opportunity structure within which transnational activism can emerge' (Tarrow 2005: 8). He identifies 'the political process that activists trigger to connect their local claims to those of others across borders and to international institutions, regimes, and processes' (2005: 11). He identifies a new stratum of activists – what he calls 'rooted cosmopolitans' – of which transnational activists are a sub-group. Transnational activists are defined as 'people and groups who are rooted in specific national contexts, but who engage in contentious political activities that involve them in transnational networks of contacts and conflicts' (Tarrow 2005: 29).

Agency is central to building movements. Leadership vision, commitment and imagination are cardinal to such projects. A gap in much of the literature is an explanation of how and why a person might become a movement activist. Such an explanation requires drawing on what could be called the social psychology of activism in order to explore how individuals are transformed from passivity to activism (Mead 1934; Fromm 1947; Cooley 1956). Drawing on qualitative interviews with activists, the core of our argument is that while global restructuring undermines agency through demoralization and depression, creating a sense of worthlessness and a corresponding lack of capacity, participation in movements transforms these self-destructive feelings, generating empowerment, creativity and a determination to resist. Harvey (2000: 237) captures the key to this psychological transformation when he poses the question, 'Where, then, is the courage of our minds to come from?' Such courage is spawned by the spirit of movement, since genuinely democratic movements assert the innate value and creativity of persons, liberating the victims of restructuring from the dungeon of their commodity status.

The labour movement problem

A third problem is whether the labour movement can be part of the construction of a counter-movement for, over the last two decades of the twentieth century, there was an almost complete consensus in the social science literature that 'labour movements were in a general and severe crisis' and that this situation has contributed to 'a crisis in the once vibrant field of labour studies' (Silver 2003: 1). Indeed, much of the emerging scholarship on social movements pays little attention to the new labour internationalism, assuming that the labour movement is a spent force – an old social movement (Castells 1997; Keck & Sikkink 1998; Tarrow 2005).

How then can the labour movement be posited as a key facet of the counter-movement when it is in a crisis produced by the very forces that need to be challenged? Part of the answer is a reinvigoration of labour studies where the discipline should not just *reflect* the decline (i.e. analyse the past); it should explore the contradictions that may create the opportunity for a counter-movement to emerge (i.e. explore future possibilities). Indeed, the crisis is beginning to produce a labour studies renaissance (Waterman & Wills 2001; Silver 2003; Burawoy 2003; Herod 2001a, 2001b, 2003; Herod & Wright 2002; Munck 2002, 2004).

The central question is whether globalization represents 'an unambiguous and unprecedented structural weakening of labour and labour movements on a world scale, bringing about a straightforward "race to the bottom" in wages and conditions', or is it 'creating objective conditions favourable for the emergence of a strong labour internationalism?' (Silver 2003: 1). Silver develops a many-sided answer. Firstly, she says, capital mobility on a global scale has undermined union bargaining power, state sovereignty, the welfare state and democracy. States that insist on maintaining expensive social compacts with their citizens risk being abandoned by investors scouring the world for the highest possible returns. The 'race to the bottom' takes the form of pressure to repeal social welfare provisions and other fetters on profit maximization within their borders. Secondly, transformations in the labour process have undermined the traditional bases of workers' bargaining power (Silver 2003: 3–5). Hyman makes a similar point when he writes that global competitive pressures have forced corporations to implement 'flexible production systems', transforming a once stable working class, replacing it by 'networks of temporary and cursory relationships with sub-contractors and temporary help agencies'. The result is a structurally disaggregated and disorganized working class, prone more to a politics of resentment than to 'traditional working class unions and leftist politics' (Hyman 1992: 62).

Silver then presents a counter-argument. Firstly, capital mobility has created new, strategically located working classes in the global South, which in turn produced powerful new labour movements in expanding mass production industries. These movements were successful in improving wages and were the 'subjects' behind the spread of democracy in the late twentieth century. Secondly, just-in-time production systems create global production chains and actually increase the vulnerability of capital to the disrupted flow of production and thus enhance workers' bargaining power, based on direct action at the point of production. She concludes, 'the more globalized the networks of production, the wider the potential geographic ramifications of disruptions, including by workers … It was only post facto – with the success of mass production unionization – that Fordism came to be seen as union strengthening rather than inherently labour weakening. Is there a chance that we are on the eve of another such post-facto shift in perspective?' (Silver 2003: 6). Thirdly, Silver suggests that the race to the bottom is the outcome of political conflict rather than inexorable economic processes undermining state sovereignty. Far from there being no alternative, assertive political struggles by labour movements have the potential to expose the idea that there are alternatives and transform the ideological environment and shift it towards more labour-friendly national, political and economic policies (Silver 2003: 7).

Silver concludes that rather than seeing globalization as a force that either strengthens or undermines the labour movement, global restructuring should be seen as a force that *simultaneously* undermines *and potentially* strengthens the movement. However, a strengthening does not occur spontaneously, neither is it 'produced' by capital accumulation alone. It requires that new sources of power be identified, a task that she begins to undertake in her path-breaking book.

The power problem

Silver (2003: 18) identifies a central problem when she says that 'the concept of "power" is largely missing from Polanyi'. For Polanyi, she suggests, 'an unregulated world market would eventually be overturned "from above" even if those below lacked effective bargaining power' (Silver 2003: 18). To overcome this lack of an emphasis on power, she utilizes Erik Olin Wright's (2000: 962) distinction between associational and structural power. Associational power is defined as 'various forms of power that result from the formation of collective organization of workers (trade unions and political parties)'. Structural power is the power that accrues to workers 'simply as a result of their location in the economic system'. Wright distinguishes two subtypes of structural power: market bargaining power which results directly from tight labour markets and workplace bargaining power resulting from 'the strategic location of a particular group of workers within a key industrial sector'.

Silver (2003: 13) makes use of this distinction to argue that market bargaining power can take several forms: the possession of scarce skills that are demanded by employers; low levels of general unemployment; and the ability of workers to pull out of the labour market entirely and survive on non-wage sources of income. Workplace bargaining power accrues to workers who are enmeshed in tightly integrated production processes where a localized work stoppage can disrupt much more widely. Different forms of structural power require different forms of organizational strategies by trade unions; in other words, different associational strategies. Thus auto-worker unions are, through the technologically interlocked assembly line, able to organize on the basis of strong workplace bargaining power. Textile workers, on the other hand, have quite limited structural power, since production can be easily rerouted, and have therefore had to compensate by strengthening their associational power (Silver 2003: 92–94).

These distinctions enable Silver to develop an innovative analysis of the new wave of campaigns to organize low-paid vulnerable service sector workers in the United States. Unions have compensated for their low level of workplace and market bargaining power by recasting associational power and developing a new model of organizing that is more community based rather than workplace based. It has also involved a more confrontational style of unionism using public tactics such as in-your-face street protests, targeting specific employers and in the case of the *Justice for Janitors* campaign, using research-intensive, lawyer-intensive and organizer-intensive resources backed by the innovative Service Employees International Union (SEIU) (Silver 2003: 110–111).

Silver's innovation can be extended further by introducing the concept of symbolic power. In her study of community based worker centres in the United States, Fine (2006: 256) identifies the importance of moral power in a context where undocumented migrants have very limited structural power. Moral power involves the struggle of 'right' against 'wrong', providing a basis for an appeal both to the public and politicians, as well as to allies in civil society. Chun (2005) uses the concept of symbolic leverage, which involves new organizational repartees drawing on the intersection between exploitation and social discrimination. In highlighting social discrimination, these repartees appeal not only to the workers who are subject to such discrimination, but also to their communities. 'Symbolic power, like structural power, is articulated with associational power, and may provide new sources of power to labour movements battling with the loss of older and more traditional sources of power in the labour market or the workplace' (Von Holdt & Webster 2006). But a call to morality, even when done in a very creative and public way, has certain limitations. In the cacophony of news events vying for public attention in the corporate media, protest action often gets lost. Activists are also drawing on what we call logistical power, a subtype of structural power. The complicated and fragile nature of the global production system gives certain types of workers power to seriously disrupt the system, and in some instances even shut it down.

Symbolic power is a subtype of associational power, since it draws on forms of social organization. But it draws it strength from taking moral claims in the workplace and articulating them as general social claims. Logistical power is a kind of structural power, since it operates in the realm of coercion. The term *logistical* is a military term and refers to the organization of supplies and stores necessary for the support of troop movements. Unlike market bargaining power and workplace bargaining power, which essentially rests on the ability of workers to withdraw from production, logistical power takes matters out of the workplace and onto the landscape where workplaces are located. It blocks roads and lines of communication, and crashes internet servers. Since work restructuring uses the politics of space to undermine structural power, logistical power turns this logic on its head. As symbolic power takes morality outside the realm of the employment relationship into the public domain, logistical power takes structural power outside the workplace and into the public domain. However, such disruptive use of power has to be combined with retaining public support, since it may actually turn citizens against the issues raised in the public domain. Furthermore, various nation-states have sought to protect corporations through anticipating this potential, thus introducing severe legal penalties for such disruptions. In chapter 9 we will argue that one of the critical tasks of labour internationalism is finding ways to circumvent these sanctions.[2]

Much of the literature credits neoliberal globalization with creating a crisis for labour movements through undermining traditional forms of bargaining power. It deepens this process by undermining marketplace bargaining power through a range of factors: the world-scale reserve army of labour; the spread of commercial agriculture which undermines non-wage sources of income; sub-contracting which undermines workplace bargaining power; the weakening of state sovereignty, which undermines associational bargaining power. Historically, associational power has been embedded in legal frameworks that guaranteed rights and through the welfare state that strengthened marketplace bargaining power. In the Second Great Transformation labour law and welfare reforms have become the centre of a strategy to roll back the power of labour. The erosion of worker power has de-legitimized trade unionism through these direct attacks and through the erosion of the welfare state. The notion that there is no alternative has had 'a powerful demobilizing impact on labour movements, puncturing a century-old belief in worker power' (Silver 2003: 16). Silver's search for new sources of power is a welcome counter to this pessimistic reading of the current situation. At this historical juncture, it seems as though the labour movement is beginning to face up to the challenge of devising strategies of its own in response to corporations' use of space and scale to disempower workers.

The scale problem

Another under-theorized area in Polanyi's work is that of scale. While the concept has a range of meanings, we refer here to the problem of linking the 'local' to the 'global', the 'micro' to the 'meso' and the 'macro'. Polanyi worked within the parameters of the nation-state, which he saw as analytically sufficient and the arena within which counter-movements evolved. In the contemporary world, however, there is a need for a more sophisticated understanding of how markets, governance and social responses are embedded in place, and how land-scapes of space and scale form the basis for contestation (Harvey 2000; Herod 2001a; Silver 2003; Tarrow 2005). The way in which many of the gains made by workers in the industrialized North during the First Great Transformation are eroded, and how workers in the underdeveloped South are kept compliant, is through corporations threatening to relocate to ever lower-wage areas. Harvey (2000: 24) refers to this exploitation of the geographic difference in the evaluation of labour as a 'spatial fix', which he defines as the attempt to resolve the internal contradictions of capital accumulation spatially. Silver (2003: 39) argues that this takes the form of 'the successive geographical relocation of capital'.

This working of space reflects the unequal power relations between global corporations and workers in civil society that is being continuously consolidated by free market globalization. Indeed, as Zygmunt Bauman (1998: 8) has argued, global corporations have enhanced their power through their relative 'independence from space' in that shareholder decision making is not tied to space in the same way that workers are place bound. Shareholders are 'the sole factor genuinely free from spatial determination' for they can 'buy any share at any stock exchange and through any broker, and the geographical nearness or distance of the company will be in all probability the least important consideration in their decision to buy or sell'. Their only concern is the maximization of short-term returns regardless of the place-based consequences of their decisions.[3] He then concludes that 'Whoever is free to run away from the locality, is free to run away from the consequences. These are the most important spoils of the victorious space war' (Bauman 1998: 9).

However, the idea of labour relegated to the 'local' and corporations as completely footloose and 'global' oversimplifies the contradictory nature of Harvey's understanding of the spatial fix. While often threatening to relocate, firms also need places where they can accumulate capital. And as Silver (2003) has pointed out, even when firms relocate, they often create new working classes and new waves of organizing and resistance. Space and scale are produced; they are contradictory and can be contested. The issue is self-evident: if corporations have increased their power immeasurably through globalizing the scale of their operations, so too can unions. 'What happens', writes Harvey (2000), 'when factories disappear or become so mobile as to make permanent organizing difficult, if not impossible? ... Under such conditions labour organizing in the traditional manner loses its geographical basis and its powers are correspondingly diminished. Alternative modes of organizing must then be constructed.' Scale is central to such an initiative, for as Herod (2003: 237) has argued, 'how we conceptualize the ways in which the world is scaled will shape how we engage with that world'. This is an ideologically charged question, for within neoliberal discourse globalization (the rationale for ceaseless corporate restructuring and spatial fixes) global scale is presented as inevitable – a scale from which there is no escape, dooming all counter-initiatives to failure. This conception views scale as a hierarchical ladder. Indeed, as Harvey (2000: 49–50) has contended, 'a hierarchy of spatial scales exists (personal, local, regional, national, global) at which a class politics must be constructed', connecting the microspace of the body with the macrospace of globalization. Herod (2003: 238) elaborates this metaphor. In this image there is a notion of moving up or down the hierarchy of spatial scales, where the global is the highest rung and where each scale is seen to be distinct. Using the metaphor of a ladder to understand how spatial scales operate has been criticized as being too structuralist and determinist. Having to conquer the local, the regional, the national, and only then the global, makes it almost impossible to conceive of a global response to corporate restructuring (Herod & Wright 2002; Herod 2003; Sadler & Fagan 2004; Gibson-Graham 2002; Latham 2002).

Instead of viewing spatial scale as a ladder, attempts have been made to use the notion of a network as an alternative metaphor. Going global does not imply scaling a ladder from the local to the global, but place-to-place linkages can form the beginnings of a global response. This understanding of how spatial scale operates resonates with notions of democratic movements, where the model is a non-hierarchical, flat, open form of a networked internationalism which maximizes grassroots participation (Herod & Wright 2002: 8). In contrast to being rendered voiceless by the domination of capital, networks have the potential to create 'spaces of hope', an opportunity for the sharing of experiences

and ideas that transcend the boundaries of place, creating newly con-
figured place-to-place relationships that have the capacity to link across
global space, which empower the local through a new sense of the
potency of global solidarity and new ideas of resistance.

However, whereas the metaphor of spatial scale as a ladder is too
structuralist, viewing it as a network – i.e. place-to-place links constitute a 'global' response – runs the danger of being too voluntaristic.
Certainly, as Herod and Wright (2002: 8) rightly point out, changing
metaphors does not necessarily change the world materially; however, theoretical conceptions may shape leadership responses, which
may lead to new organizational initiatives. So as to fully capture the
agency/structure problematic, we contend that the network metaphor of scale is the most fertile conception (political/organizational
choice) for building a place-based global social movement response
to corporate restructuring. However, we feel it is important to conceptualize a dynamic interplay between place-to-place global networking
and the different levels of the consolidation of corporate and political
power – local, regional, national and global. Each is a potential and
indeed necessary field of struggle, through which movement power
may coalesce. Conceiving networked scale outside of these terrains of
struggle is limiting. This conception of the interplay between networks
and levels of power provides a more nuanced understanding of how
spatial scale operates and how power struggles between capital and
labour may unfold spatially.

In this new connectedness across space, a point of continuing debate
is the need to engage the state in the struggle against the spatial fixes
of global corporations. As Rutherford and Gertler (2002) have argued,
the nation-state as a scale of contestation does not disappear under
neoliberal globalization. Often, claims are still made at the level of the
nation-state, or even sub-state governmental institutions such as provinces, districts or cities. As Harvey argues, 'the left must learn to fight
capital at both [or, one might add, multiple] spatial scales simultaneously. But, in so doing, it must also learn to coordinate potentially
contradictory politics within itself at the different spatial scales' (Harvey
2001: 391).

The dilemma of integrating different struggles at various spatial
scales is the central challenge in constructing a new labour internationalism. As we argue in chapter 9, the key obstacle to such a task is that
globalization increases workers' sense of insecurity, with the potential
to turn them inwards on their private troubles, thereby undermining active agency. In the absence of agency, a new kind of global

counter-movement response to corporate spatial fixes that responds proactively through place-to-place global networking and through struggles against national states that protect corporate interests against citizens will be a pipedream, stillborn at the moment of conception. Whether we view spatial scale as a ladder or a network, or even a dynamic interaction between the two notions, rebuilding place-based social movements that are profoundly democratic and participatory is the key to the new civil society response of working space through networking scale. There now exists a number of practical experiments along these lines, which we will detail in chapter 9. We also consider the opportunities and the obstacles to such a venture. These represent the first signs of the emergence of a global unionism as a core constituent of a global counter-movement.

However, we do not, in this book, provide blueprints of how a counter-movement could be constructed. Instead, we begin the first step in such a project by grounding our analysis in the everyday lives of workers, their households and their communities in three places: Ezakheni, Changwon and Orange. It is this triangulation across three research sites that enables us to ground our theoretical and political narrative in empirical case studies. This sets our book apart from the numerous studies on globalization and labour that have been written over the past decade, which are largely based upon assertion, rather than detailed evidence of change on the ground.

We have suggested that, in the current Second Great Transformation, Polanyi's problematic poses a range of problems of its own. These relate to how we understand society, the fact that counter-movements do not arise spontaneously, the role of the labour movement in such a counter-movement, the sources of power that movements can draw on, and the need for solutions to be sought at scales that range from the local to the global. We turn now to our research strategy.

Researching Working Life

We have chosen three very different places to illustrate how workers' lives are the product of struggles between contending social forces, struggles that lead to different relationships between the market, the state and civil society. The way they engage with globalization is also shaped by their physical geography. South Africa has been inserted into the global economy through its rich mineral resources, but remains relatively remote from the new global markets. Australia, on the other hand, is surrounded by the new competitive markets of East Asia and this has shaped its radical restructuring of what was a highly protected market. Korea, by contrast with South Africa and Australia, has limited natural resources, and has found its niche in the global market through direct state intervention and labour repression. This has enabled it to build a powerful manufacturing industry. Each society has emerged over time, we suggest, through the social relationships men and women have constructed in order to produce and appropriate wealth. For us, there is no Chinese wall separating the sphere of reproduction (the household) from that of production (the employment relationship); work – paid or unpaid – takes place in both spheres. Indeed, they are intimately and asymmetrically connected! This is why we have approached the analysis of society by looking at the workplace, households and communities. We draw on world-systems theory, particularly Wallerstein and Smith (1992), to explain this relationship by defining the household as a socially constructed unit moulded by the changing patterns of the world economy.

But unlike world-systems theory, we foreground the self-activity of workers, in particular organized labour, as it is workers who contest the nature of the labour process and how its rewards are distributed between capital and labour. History, and the distinctive developmental paths these countries have taken, is crucial if we are to understand the different ways in which these relationships have been constructed in all three places. We argue that global restructuring is impacting negatively on the lives of working people, as Barbara Pocock (2006) dramatically demonstrates in her book, *The Labour Market Ate My Babies*. Under the impact of global competition, workplaces are transferring the stresses of the workplace onto the household, creating conflict within the home, such as spousal abuse, abuse of the elderly, substance abuse and abuse of children. We call this a crisis of social reproduction.

We chose Ezakheni and Orange as our research sites for exploring this crisis as we had been conducting research on workplace restructuring and its impacts on households in these two places for five years prior to this project. Between 2002 and 2006 we conducted research on the white goods factory in Ezakheni and in Orange. We have drawn on the earlier research in this study, giving it a historical perspective.

The central source of data for this study was a survey conducted between September and November 2005 in all three places using the same semi-structured interview schedule. We identified respondents through using a snowball sampling technique, while attempting to ensure a spread of workers who worked in different parts of the factories and on different contracts of employment. The survey was complemented by simultaneous in-depth interviews with key actors: trade union leaders, local government officials, community workers and activists. These interviews are listed in the bibliography along with the interviews we have conducted on labour internationalism.

An additional part of our research strategy was ongoing observation of working life in the three sites as well as intensive participant observation in SIGTUR over a six-year period. An important part of our research strategy is our engagement with the local union structures through ongoing report-backs, both formal and informal, at different stages of the research. This provided us with an opportunity to share our research findings, but also to test the accuracy of our data and the coherence of the argument we were developing as it evolved. Of course, we used a range of primary documentary sources, including the internet, covering company annual reports and other documents, newspaper reports and government documents, as well as reports drawing on existing data generated by census surveys. We also consulted the key secondary texts on labour, politics and society in all three of the countries.

Comparing three towns in three different countries raises a number of conceptual, theoretical and methodological challenges. The first is how to obtain the same depth of information in all three research sites. Inevitably our knowledge of Australia and South Africa is deeper, but we hope we have overcome some of these limitations by engaging a South Korean researcher for our interviews and data analysis on this country. We are also limited by the fact that we do not speak or read Korean, and had to rely extensively on translation of the written and spoken word. A similar limitation applies to South Africa where our respondents were largely Zulu speaking. Fortunately, they all spoke English and we used a Zulu-speaking research assistant to conduct some of the research in Ezakheni.

A more fundamental challenge in comparative social science is the relationship between similarity and difference. By taking the same production process – the manufacturing of refrigerators – and investigating the impact of globalization on the companies that produce these refrigerators, we found a great deal of similarity between the three workplaces. However, our aim was not simply to examine the impact of restructuring on the workplace; a central aim was the impact of restructuring on the sphere of reproduction, on the households and communities where women and men live. This is why we chose three places and entered the 'hidden abode of reproduction'.

Whereas the workplace revealed similarity, the households and communities revealed variation, variations that are too great to be explained by institutional differences in the nature of their industrial relations systems. To explain these differences it became necessary to examine the socio-political context in which these labour market institutions are embedded, and the forms of organization that emerged in response to these institutions in all three countries. By adopting this approach different challenges emerged in the three different countries. The most striking differences arise out of the legacies of colonialism and under-development in South Africa, the creation of an advanced industrial society in Australia, and the transformation of Korea over a 40-year period from an economically backward country to a successful industrialized nation.

This approach, what could be called a 'contextual comparison', differs from the conventional political science method of 'matched comparisons' which compares countries facing similar global developments (apples with apples). Instead, contextual comparisons encourage comparison between different challenges across different countries. This is a difficult task as, to pursue the metaphor, it involves comparing apples with oranges (Locke & Thelen 1995: 338). Such a task, we believe, requires an examination of the historical evolution of the three societies, a task we undertake in chapter 8.

The book is divided into three sections. Part One, *Markets Against Society*, demonstrates how workers in all three places are experiencing a growing sense of insecurity as the employment relationship is reconfigured under the threat of global hyper-competition. We draw on Burawoy's pioneering work on comparative labour regimes to explain the fluctuations in workplace regimes, in particular his use of Antonio Gramsci's concept of hegemony to explain the ideological foundations of worker compliance (Burawoy 1979, 1985). We show how liberalization has resulted in a growing concentration and rationalization in the white goods industry and how the neoliberal paradigm is leading to the privatization of essential services in all three countries.

Part Two, *Society Against Markets*, examines how workers in all three places are responding to these pressures. We identify two types of responses, both individual and collective, formal and informal, to rapid market liberalization: on the one hand retreat from, and adaptation to, the market, and on the other, mobilization against the power of the market. In order to examine the impact of restructuring on workers' lives, we examine the structure of the households and the nature of the communities in which they are embedded.

In Part Three, *Society Governing the Market?*, we delve into the histories of these three countries to explain why each society is responding differently to liberalization. We suggest in chapter 8 that these trajectories explain the substantial differences that exist across the three cases with respect to the ways in which workers, union leaders and political parties have responded to neoliberal globalization. We then examine new forms of transnational activism and the growth of a new labour internationalism. We suggest that the problems we identify in these three countries cannot be resolved at the level of the nation-state alone. The challenge, we suggest, is to imagine a real alternative, what Erik Wright calls a real utopia, grounded in the actual experiments and institutional forms that are emerging, which link the local to the global. We conclude by identifying the conditions necessary for a successful counter-movement. The first condition is the existence of a convincing critique of the existing social structures demonstrating how these structures impact negatively on working life. This we do in Parts One and Two. The second condition is the existence of a feasible alternative to the current structures. A final condition is a realistic map of how to attain this alternative. We begin tentatively in Part Three to address these two conditions, but we make it clear that this is an activity that goes beyond the capacity of scholars alone; it requires a partnership between engaged intellectuals and those institutions and movements that have begun to realize that a democratic alternative to the socially disruptive impact of unregulated markets is possible.

We turn now to Part One for an examination of the changing workplace.

Part I

Markets Against Society

We begin our analysis by entering the 'hidden abode of production'. Here, real commodities – refrigerators, washing machines, cookers, microwave ovens, those taken-for-granted appliances that shape our everyday lives – are produced. As in all production, technological revolutions impact on the household. Cooking was initially done on a wood fire. Later, coal was used and iron and steel foundries produced the cast-iron stoves. In both Australia and South Africa the white goods industry grew out of these foundries. Today, cooking in the three countries we study is predominantly done on electric cookers, although most rural and many black urban households in South Africa still rely on wood, coal and paraffin.

The preservation of food has always been a central concern in the household. In Korea, one solution to preserving food was to pickle and ferment vegetables. These are the origins of *kimchi* (alternatively, spelled *gimchi*), which is so central to Korean culture that a museum dedicated to the history and meaning of *kimchi* was established in Seoul. *Kimchi* is a bit like German sauerkraut, but key ingredients are garlic and chilli peppers. These ingredients do not originally come from Korea though. Chilli comes from the Americas and was only introduced to Korean cuisine in the eighteenth century. This shows how the first globalization – the spice trade – historically shaped culture. Indeed, local cultures have been spiced by the global for centuries. Nowadays much of the *kimchi* consumed by Koreans is mass produced, most of it in China. Koreans now have specialized *kimchi* refrigerators to store this. LG is one of the producers of these *kimchi* refrigerators – a product unique to Korea (Lee 1991).[1]

In the 1970s and 1980s, labour process theory revitalized an interest in the world of work. Like the automobile, household appliances are central to consumer capitalism. In Australia the post-Second World War manufacturing boom created the 'Australian lifestyle', of which the refrigerator was one of the key symbols. In such a hot, dry climate, the cold beer ('stubbie') in the fridge reached almost iconic status among mates. Similarly, white goods are as central to South African consumer culture. The name of Defy's washing machine, the Automaid, however, takes on a somewhat ironic meaning when one considers the organization of household labour around cheap, black, domestic 'maids' who cleaned, cooked and cared for the children in the sprawling white suburbs.

In the 'post-modern', 'post-industrial' world of the 1990s, the realm of consumption and consumer culture dominated discourse and even scholarship. Factories were relegated to the past. However, with the globalization of production, and those taken-for-granted factories closing down in the industrialized North, suddenly, the study of work has re-emerged as a central focus shaping the human condition and the popular imaginary.

To understand the dynamic of change in the modern workplace it is necessary to examine transformations in the structure of production characterized by a high level of concentration reflected in the domination of sectors by only a handful of large global corporations. The growth in the size and power of global corporations, Leys (2001) argues, gives them the capacity to transform politics and the state. He identifies three changes driving the transformation of politics. Firstly, deregulation and the creation of global markets has accelerated the process of concentration to the point where every sector is now dominated by ten to twelve large corporations, and sometimes even fewer. This is reflected in the fact that the measure of the degree of concentration has become world market share rather than national indicators. Secondly, global corporations have increased in number from 7,000 in 1970 to 60,000 by the 1990s, creating tightly integrated production systems and marketing strategies. Thirdly, the leading corporations have considerable financial leverage. In 1997 the largest 100 corporations had total sales of $39 billion and assets of $42 billion, which meant that they had negotiating resources that outstripped those of nation-states. These companies invest enormously in information technology, giving them a marked communications advantage over states. As a result they are powerful lobbyists with instant access to government leaders. The white goods industry is no exception with regard to these changes, as we illustrate in chapter 2.

In this context, Leys's (2001) work is important, for it highlights the erosion of democracy and citizenship produced by the power shift from civil society to corporations (Leys 2001: 6). Such a transformation is the defining characteristic of market-driven politics, which captures Polanyi's concern for society when market forces (corporations) escape political control. This change does not reflect a weak state but rather a captive state, a politically transformed state in which the governing party *chooses* to embrace the agenda of corporations against the will of civil society. Here the role of the state has changed from protector of the vulnerable to defender of global corporate interests.

How do these global processes of corporate restructuring and market-driven politics play out in diverse local contexts? What are the possibilities of resisting the destructive elements of global restructuring? These are the questions we set out to answer by using one industry, but three different firms in three very different countries, as our vantage point. In chapter 2 we describe how the rapid liberalization of markets in Australia and South Africa is eroding the local manufacturing industry as mining and finance capital reassert their hegemony. South Korea, in contrast, has succeeded in developing a highly successful trans-national white goods company, LG, a product of its developmental state. The key role of the state in shaping the labour system in Korea will be discussed in chapter 8.

In chapter 3 we show how the nature of control of labour has converged in all three countries towards growing flexibility in the work-place, leading to a regime of control that can be characterized as market despotism. At the centre of this shift is a change in the nature of the employment relationship from full-time, indefinite employment, to various forms of precarious fixed-term, part-time and outsourced employment. This shift has been accompanied by an intensification of work and growing insecurity. The liberalization of markets and investment has enabled capital to globalize production, taking advantage of the uneven nature of capitalist development to relocate to low-wage areas of the globe. This convergence to an 'old' form of control, market despotism, has been accompanied by the withdrawal of the policies that protected local white goods industries and created relatively secure employment. Workers are constantly made aware of the fact that their places of work may not be around much longer if they are not compliant.

In chapter 4 we examine how the mantra of the market asserts its global reach through the privatization of essential services. Importantly, the attempt to privatize the electricity utility in Korea was pre-empted by resistance from organised labour. We describe a similar process of the commoditization of essential services in Australia and South Africa. While in Australia the privatization of electricity in the State of Victoria met little resistance, the privatization of water in partnership with global and local companies has led to widespread but sporadic resistance at the local level in South Africa. An important part of our approach is to go beyond the insecurity created in the workplace to an examination of how the privatization of essential services, a key part of what Harvey (2003) calls accumulation by dispossession, is creating insecurity in the household.

2

Manufacturing Matters

In this chapter we examine the twentieth-century rise of white goods production in Australia, South Africa and Korea, and show how recent global restructuring accelerates the growing dominance of a small number of powerful global corporations, a phenomenon which has vast implications for locally based production. Indeed, the strategy of the global corporation is a 'profoundly geographic affair' (Harvey 2000: 57) characterized by scant commitment to place, to the societies within which they ground their operations, for these shareholder-driven entities are ever watchful, restlessly scanning the globe for more profitable production sites where taxes and wages are lower and subsidies (incentives) higher. In this chapter we analyse the impact of this transformation on the white goods corporations which emerged in the three countries we consider.

The comparisons are fascinating since they tell a story of the geography of production, that is, how global corporations work space to consolidate their power and how the changing role of the state is so central in this transformation. The comparison reveals how the local impacts of this global concentration of corporate power in the hands of a few mega corporations differ markedly. The Australian company Email became the takeover target of the Swedish company Electrolux, the second largest global corporation in the sector. The acquisition was effected in late 2000, confirming the trend of overwhelming foreign ownership in the Australian economy (Crough and Wheelright 1982). Despite facing similar pressures, Defy the local company in South Africa has resisted this global logic and the company remains a South African owned entity which still dominates the South African market, but is experiencing the pressure of global markets and imports.[1] When we bring Korea into the equation we witness a quite different trajectory with a marked contrast in outcomes. We track the remarkable story of LG, which became a global player in a matter of decades, successfully challenging the giants on the world stage such as Electrolux and Whirlpool. We develop this analysis which traverses brief histories of the development of these three factories and the ways in which they have responded to the global restructuring of the sector because this enables us to ground and explain the changes in the organization of work which we consider in chapter 3. There is an important link here. The restructuring of work and the insecurity it has produced cannot be viewed in isolation. Highly localized changes in the way work is organized are driven by the deregulation of national economies, which has advantaged the large global corporations against the local producers. While Australia and South Africa simply stripped away their tariff controls and hoped for the best, the Korean regime pursued an alternative strategy that used state power to leverage advantage for Korean corporations on global markets. The fact that LG is now a leading global player has to say something about the veracity of this counter-market strategy and the vacuous alternative of the South African and Australian governments.

We begin by analysing the emergence of the sector in each country, revealing the cardinal role of the state in securing the manufacturing of white goods. We then provide an overview of the concentration of production at the global level. This forms the backdrop to the radical shift in industry policy in South Africa and Australia as the role of the state is transformed from the guardian of local production to the handmaiden of global interests as free market ideology is embraced without restraint. Corporate restructuring accelerates as a consequence, often with destructive economic and social impacts at the local level.

Before moving to this analysis, we need to make a significant point. Our analysis of the local impacts of the spatial fixes of global corporations leads us to conclusions at odds with the market model of globalization which contends that the geographic shifts in production reflect the law of comparative advantage in action where a loss in manufacturing in one country is balanced by new developments in its service sector, hence the global economy needs to be viewed in the totality of its specialized economic functions. In contrast, we contend manufacturing matters because of 'direct linkage: a substantial core of service employment is tightly tied to manufacturing' (Cohen & Zysman 1987: 3). In their critique of the post-industrial service society, the authors argue that manufacturing complements the service sector and can never be substituted by the growth of services. 'Were America to lose mastery and control of manufacturing, vast numbers of service jobs would be relocated after a few short rounds of product and process innovation, largely to destinations outside the United States, and real wages in all service activities would fall, impoverishing the nation' (Cohen & Zysman 1987: 7). Although the argument concerns the economy of the United States, the authors are making the more general point that a strong manufacturing sector provides the foundations upon which a secure society is built, where the geography of production reflects an even spread.

Building the Nation

What is common in each country is a historic commitment to manufacturing as the means to establish a secure society and a prosperous, independent nation. In the case of South Africa and South Korea which fought independence battles from colonial masters, this was acute. The state was the key instrument in realizing these goals. For Australia, a resilient manufacturing base provided a material base to build a social democracy where equity and fairness were advanced as core cultural values (Pusey 1991).

Australia

In Australia a commitment to constructing social democracy emerged at the turn of the nineteenth century. It was grounded in fostering and protecting a diverse local manufacturing sector that would create secure jobs and wages that met human need. Promoting a local white goods industry was a facet of this venture. Stoves, refrigerators and washing machines were first manufactured in the 1920s, stimulated in no small measure by a protectionist state committed to manufacturing (Clark 1983: 29). If the expansion of production is a measure of value, this was a golden era as some forty factories came into existence, thriving on industry assistance and infrastructure development that reduced risk and encouraged local investment (Clarke 1983: 35). This was a nation-building state where economic development was shaped by defined social goals. The tariff regime, established after Federation in 1902, shielded Australian white goods manufacturing from cheaper commodities produced elsewhere.

The tariff strategy reflected the nature of settler capitalism (Denoon 1979) and Australia's intermediate position in the world economy.

Crough and Wheelright (1982: 88) argue that 'on the criteria of economic efficiency, Australia would most probably be almost exclusively an agricultural and mineral producer, with a smaller population'. The authors point out that smaller capitalist economic systems such as Australia have only been able to survive through the creation of forms of state capitalism, which have to a large extent protected both capitalists and workers from the 'inroads of world competition by those bigger, stronger and more efficient' (Crough & Wheelright 1982: 91). The role of the state was to regulate competition and protect national companies from predatory global corporations. Regulating the economy in this way was integral to 'mastering risk' (White 1992). Globally, Australia's position is relatively weak due to the small size and remoteness of its market, hence strategies needed to evolve that master what White (1992: 270) refers to as 'power risk'. Such risk arises from two situations of subordination: firstly, the position of the Australian state relative to other states and, secondly, the relationship between Australian companies and global corporations.

Tariffs were central to this 'mastering of power risk'. Historically, a Tariff Board was created as the principal instrument of a state partnership committed to the development of a diversified manufacturing sector in Australia. The Brigden Committee noted, 'The diversion of production to the protected industries has increased the diversity of occupations and opportunities, and introduced more stability in the national income than if it had been more dependent on the seasons and the vagaries of overseas markets.' The tariff 'was based on a social vision, and gave rise to a very distinctive society' (quoted in Crough & Wheelright 1982: 88). This mastering of risk aimed at minimizing insecurity.

Within this policy context, Kelvinator Australia and Simpson and Pope Industries emerged as significant white goods corporations, while Email, set up in New South Wales (NSW) in 1934, became Australia's leading producer, expanding its workforce from 400 in 1946 to 2,150 by 1974. In 1946 the Commonwealth government's small arms factory at the country town of Orange was leased to Email, which developed the facilities into a large refrigeration plant. Economic development was not left to market forces. The state had specific social goals that it sought to realize that required intervention. Thus the NSW government subsidized the costs of corporations that were willing to decentralize production and thereby contribute to the growth of regional economies, offering freight concessions and financial incentives. The government also expanded its housing programme in Orange, making the site attractive to skilled tradespersons at a time of acute labour shortage

(Linge 1963; Boyce 1976; Jacka & Game 1980; Clark 1983: 33). Email also received loans and payroll tax rebates (Loveday 1978). As a result of these interventions, the company became one of Australia's most significant white goods manufacturers during the postwar period. These synergies between state and local corporations that created a reasonable degree of predictability in the lives of citizens were transformed after 2002. But before we tell that story, we analyse the emergence of white goods in South Africa.

South Africa

The emergence of South African white goods manufacturing was shaped by the country's geographic location, which afforded some degree of natural protection, a phenomenon reinforced by a tariff policy of import substitution. In this regard, the trajectory appears similar to Australia. However, a major difference is the dissimilar paths these settler colonies took regarding internal labour regimes. While Australia depended upon a policy of white immigration to address problems of labour supply, South Africa relied on the availability of abundant, cheap black labour. Indeed, the colonial state played a key role in securing labour supply for industry. Whereas Australia created a welfare state for all citizens, South Africa moulded a dual society, with social protection for the white settler population, which excluded Africans and other persons of 'colour'. Some (e.g. Gelb 1991) refer to this as racial Fordism, since the producers of consumer goods were primarily black and the consumers were primarily white. Such a strategy also limited the size of South African consumer markets, and industry never really developed the economies of scale characteristic of Northern Fordism.

A second difference is the way in which industry was shaped by the bifurcated apartheid state. By this we mean the stark urban-rural divisions on which the apartheid system tried to create 'independent homelands' for Africans in order to legitimize their political exclusion from citizenship rights in the rest of South Africa. This created a geography that could be exploited for other forms of control, including labour control, as will be described in chapter 3. This internal uneven development, along with different political regimes in urban and rural areas, was developed as an internal spatial fix to worker militancy. When semi-skilled black workers organized themselves into trade unions in the 1970s, companies, assisted by the state, relocated production to rural areas that were part of apartheid's 'homelands' – pseudo-independent tribal regimes (Mamdani 1996; Hart 2002; Bezuidenhout 2004, 2006).

Ways of facilitating the growth of local manufacturing were first considered in 1910, shortly after four British colonies joined to form the Union of South Africa. In 1911 the Iron, Concrete and Asbestos Manufacturing Co., Defy's early predecessor, unsuccessfully petitioned the Trade and Industries Commission to increase tariffs on imported goods (Rosenthal n.d.: 23–25). However, the colonial state was not ready to protect local industry since an increase in the cost of wage goods would impact negatively on the dominant mining industry (Kaplan 1977: 180). The first major growth spurt in manufacturing was stimulated by the First World War. When foundries in the United Kingdom turned to manufacturing weapons and ammunition, their South African counterparts stepped in to fill the market gap. The number of factories in the country as a whole increased from 3,998 to 6,890 from the period 1915–16 to 1919–20, and employment increased by 73 per cent. As soon as the war ended, manufacturing profitability slumped when local manufacturers faced competition from importers once again (Kaplan 1977: 180–181).

In 1924 the government of Jan Smuts was defeated at the polls by an election pact between the Afrikaner 'nationalist', Barry Hertzog, and a party representing the interests of 'white' labour. Their election manifesto unequivocally declared 'a definite policy of protection for industry'. Consequentially, the Customs Tariff Act of 1925 introduced general protection, while also privileging white labour. This was indeed a nation-building project, but one which consolidated white privilege.

In the context of the protectionist regime of this Pact Government, a local white goods industry was consolidated and brands such as Univa, Ocean, Defy and Barlows later became household names. Some brands were manufactured under licence, while others were developed locally (see Baumann 1995). It was during the Great Depression that Durban Falkirk turned out its first electrical appliances. In 1932 simple hot plates were manufactured and later, kitchen stoves. Soon after the introduction of electrical appliances, the Durban Falkirk company introduced 'Defy' as a brand name – by the late 1950s, 'electric ranges', 'hot plates', 'radiators' and 'compound boilers' were produced under this brand (Rosenthal n.d.: 51–57).

In 1948 the Afrikaner National Party (NP) came to power. Its industrial policies built on the existing structure, but they further developed racially exploitative labour market controls through an extension of pass laws and labour bureaux. The protectionist regime was bolstered by the introduction of exchange and import control measures in order to strengthen local manufacturing (Davies et al. 1976: 27; O'Meara 1982).

With South Africa's black population firmly controlled by Afrikaner Nationalists, white South African consumerism prospered to the jingle of 'Braaivleis [the Afrikaans word for barbeque], rugby, sunshine and Chevrolet'. In this climate the local white goods industry thrived.

Resistance to the stricter forms of 'influx control' and political repression by the regime led to mass resistance, with the massacre at Sharpeville in 1960 becoming a catalyst for future resistance. This event led to mass capital flight, and South African firms stepped in to buy 'a very wide range of industrial undertakings' cheaply (Davies et al. 1976: 28). Capital flight by foreign companies led to a period of monopolization of the South African economy and so Defy also ended up in the hands of Sankorp, the industrial arm of Sanlam, one of the major Afrikaner financial powerhouses created by 'Volkskapitalisme' (see O'Meara 1982). With the rise of the militant industrial unions in urban areas, the state also supported capital's spatial fix to rural areas under the control of pseudo-independent tribal authorities. The official term for this strategy was 'industrial decentralization'. In 1986 Defy set up its new refrigeration plant in Ezakheni, one of these decentralization zones.

By the end of formal apartheid in 1994, South Africa had a significant white goods manufacturing industry that employed a substantial work-force in places like Johannesburg, Durban, Ezakheni, Isithebe and East London. However, a process of concentration had already begun. The end of apartheid would lead to restructuring and plant closures. But first we show how Korea was able to build one of the world's top ten white goods companies. They were in league with the big players, while Email and Defy struggled and floundered.

South Korea

The widely proclaimed mantra of the architects of the world economy is the notion that freeing markets will stimulate competition on a global scale and as corporations compete the efficient will prevail to the benefit of consumers. South African and Australian political leaders who unquestioningly bought into this vision have refused to concede that states do alter the terms and conditions of global competition. Korea is a fine illustration of the shallow, flawed nature of this free market pro-position, for companies such as LG were nurtured by the authoritarian, militarized character of the Korean state, which sought to guarantee a cheap, disciplined labour force.

In contrast to Defy in South Africa, or Email in Australia, which never transcended the local market, the LG corporation in Korea has become one of the world's most successful global corporations. In fact it is one of the ten largest corporations dominating the world white goods market. The success story of LG is grounded in the militarized character of the Korean state and the strategic choices it made in the postwar period. When a military regime was established under the leadership of Park Chung Hee after a coup in 1961, a strategic choice was made to emulate the Japanese experience of the Meiji period. Park set about identifying business leaders who could develop Korean companies along the lines of the Japanese Zaibatsu. He aimed to create a state capitalism, what Amsden (1989) refers to as the 'developmental state'.

The rapid growth of large diversified companies (chaebols) along the lines of Japanese corporations was generated through a high degree of state intervention. The military nationalized commercial banks, which gave them a powerful instrument to advance corporate interests. Chaebols received huge tax breaks and massive subsidies, and they were allowed to sell on the domestic market at inflated prices. The military regime sought to tightly control labour to advantage the aspiring Chaebols. All these facets of state intervention coalesced to advantage Korean companies on global markets.

LG, the successful global player of today, is a product of this strategy. LG, which stands for Lucky Goldstar, grew out of the Lucky Chemical Industrial Company, which was founded before the Korean War, in 1947, as a partnership between the Ku and Heo families. It was the first chemical company in Korea. Historically, the two families had a business relationship. Ku In-hoe, who founded the firm, came from the older Korean elite. His father was a landowner who diversified into textiles during the period of colonialism under Japan (Cumings 1997: 329). When the Korean War ended in 1953 with all the social dislocation and disruption produced by such events, the military emerged as the strongest institution. National conscription meant a considerable part of the population was socialized into a militaristic form of discipline and this was transposed into corporate culture. President Syngman Rhee used the postwar period to extract maximum aid from the US, and set up policies to substitute Japanese industries for Korean (Cumings 1997: 302–310). Local businesses benefited immensely from these policies, and in 1958 the Ku and Heo families launched the Goldstar Electronics Co., Korea's first electronics company. 'Lucky' became their brand name for household chemicals such as washing powder and toothpaste. These products were not exported. Goldstar became synonymous with consumer electronics, and became a major exporter of such goods. In 1959 they produced Korea's first radio.

After the 1961 coup, Park Chung Hee committed to assisting companies such as Samsung and Goldstar, who were successful in the by-now saturated local market, to focus on developing export markets with government assistance (Cumings 1997: 311–312). Goldstar exported its radios from 1962 onwards to the USA and Hong Kong. In 1965 Goldstar produced Korea's first refrigerator and in 1966 its first black and white television set. In 1968 air conditioners followed and in 1969 Korea's first washing machine was manufactured. The economic 'miracle' was about to take off.

In 1973 General Park announced that six specific industries would be targeted for export growth. These were steel, chemicals, automobiles, shipbuilding, machine tools and electronics. The southwestern region, incidentally where most of the military elite came from, became a hub of manufacturing, when Pohang was developed into a major industrial port. Changwon and its port Masan became a hub of engineering activity (Cumings 1997: 322–326). In 1977 Goldstar started to produce colour televisions. By 1978 the corporation's exports exceeded US$100 million, a first for the South Korean electronics industry. But the increasingly authoritarian military regime also clamped down on human rights, declaring all work stoppages illegal by decree in 1973 and making criticism of the regime a violation of national security in 1974 (Cumings 1997: 358; Chang 2002).

In 1979 there was a sudden downturn in the Korean economy. The country recorded a negative growth rate of 5 per cent and lost 6 per cent of its GNP the following year. The government also violently suppressed protest action by female workers at the YH Trading Company, a textiles firm, as discussed in chapter 8. Massive protests spread throughout the country, which shocked the military regime. Park Chung Hee was assassinated by the chief of his secret intelligence service, and Major General Chun Doo Hwan took power. Pro-democracy protests in the town of Kwangju were violently put down by the state, which became even more authoritarian and unpopular than the regime of Park Chung Hee. By 1983 the economy had started to recover from the temporary downturn (Cumings 1997: 367–382). Goldstar continued to expand its overseas operations in the early 1980s. In 1980 its first European sales subsidiary was established in Germany, and in 1982 the corporation established a colour television plant in Huntsville in the USA. By 1984 its sales had surpassed one trillion won. In 1986 a VCR plant was established in Germany.

During Chun's term students, workers and religious groupings continued to organize against the regime, often sending students into factories as 'disguised workers' to organize labour unions. Uprisings in the late 1980s culminated in elections in 1993, formally ending military rule (Cumings 1997: 386–388; Chang 2002). Nevertheless, a legacy of the military regime was a strong and diversified local manufacturing industry.

The significance of the LG success story for our comparative analysis is the light it sheds on the *character* of global competition. Two factors interlocked, giving LG a significant edge on global markets. Unlike South Africa and Australia, the military state in Korea firstly protected LG's local market share, while empowering it through economic measures to secure an increasing share of global product markets. Secondly, the repression of labour sought to ensure advantage both through labour's comparative cheapness and its military-like discipline.

The Korean experience, and indeed the experiences of many other Asian 'developmental states', necessitates that one treat with scepticism the claims of institutions such as the IMF and the World Bank that stimulating competition by deregulating markets will lead to development. The Korean case, as illustrated by LG's ability to become a global player in white goods manufacturing, has unfortunately not been taken seriously enough by South African and Australian political leaders. Instead, they assumed that rapid trade liberalization would facilitate higher levels of productivity in their manufacturing sectors. The demise of the Electrolux factory in Orange and the inability of Defy to become a global player should be seen in this light. Polanyi's enduring insight is the way an almost mystical notion of the market obliterates all other considerations, such as for example the implications of open market competition between democratic states with free trade union rights and military regimes which suppress these rights. Later in this chapter the consequences of these lacunae become apparent. However, before venturing into these effects, the actual nature of global competition which is so threatening to local manufacturing will be further clarified.

The Emergence of the Global Mega Corporation in White Goods

We have described how the state in all three of the countries in this study established a viable, but protected, manufacturing industry during the course of the twentieth century. However, at the end of the twentieth century this sheltered world experienced a dramatic challenge as major spatial shifts began to occur in white goods manufacturing. One of the reasons for this was the saturation of the markets of Europe and the United States, as well as the increased importance of the emerging markets of Brazil and China. Global trade liberalization had enabled firms to rationalize production by creating mega factories to serve emerging markets as well as their more traditional customers in the economic core. As a result, Brazil, China, Mexico, Russia, India and Turkey emerged as major hubs for manufacturers (Nichols & Cam 2005). These corporations used think-tanks to politically advocate free trade to ensure that this global dynamic was feasible (Cockett 1994).

The white goods industry essentially competes on the basis of volume and is highly concentrated. As table 2.1 shows, the five largest corporations controlled 30 per cent of the market with a combined turnover of US$45 billion in domestic appliance revenues in 2002. The top two white goods companies alone had 15 per cent of global volume sales, while the ten largest corporations account for 44 per cent of the global market. The table shows that the developed world's global corporations still dominate, although LG is now in the pack. The five North American and European MNCs have secured just over one third of world sales (35.6 per cent). The four Japanese MNCs achieved 14 per cent of world sales, with the emergent MNCs from China and Korea rapidly growing their market share.

Table 2.1 Ten largest white goods corporations (as measured by share of total global volume sales) 2001 and 2002

		% volume	
		2001	*2002*
Whirlpool	(US)	7.9	7.9
Electrolux	(Sweden)	7.3	7.1
Bosch-Siemens Hausgeräte	(Germany)	5.8	5.7
General Electric (Appliances)	(US)	5.3	5.4
Haier Group	(China)	3.2	3.8
Matsushita	(Japan)	3.1	3.2
Maytag	(US)	3.0	3.1
LG Group	(Korea)	2.4	2.6
Sharp	(Japan)	2.6	2.6
Merloni	(Italy)	2.3	2.5

Source: Euromonitor: Global Market Information Database, 'The world market for domestic electrical appliances', November 2003

This high degree of concentration is the outcome of a process of capital accumulation driven by intense competition between private corporations where 'one capitalist always strikes down many others' (Marx 1976a: 929). Competitive war between private companies is the 'driving fire' of the rationalization of production (Marx 1976b: 254), which results in the expropriation of the many by the few, leading to an extreme concentration of corporate power. In recent decades this has further accelerated as a result of the economic and financial deregulation neoliberal globalization carries in its wake. Such a power concentration enables the biggest to become ever more proactive in the takeover wars, which then generates further rationalizations. So, for example, in 1982 there were 350 corporations producing white goods in Europe. A mere decade later, this was slashed by two-thirds to 100 companies. By the mid to late 1990s, a mere 15 companies controlled 80 per cent of the European market (Segal-Horn, Asch & Suneja 1998: 105). Globally, the competitive war between the major players is being waged through spatial fixes and lean production restructuring. Also, new players from China (Haier) and Turkey (Archelik) are emerging to take on more established multinationals such as General Electric, Whirlpool, Electrolux and Siemens-Bosch (Goldstein, Bonaglia & Mathews 2006). We now turn to a consideration of the impact this restructuring has had on the three factories we researched.

The Dynamic and Impact of Global Restructuring

Electrolux in Australia

When global corporations in the North first experienced market saturation in the advanced Northern economies during the 1970s, as demonstrated above, they responded by pressurizing governments in the South to deregulate and allow greater market and investment access. This was an influence in the seismic shift in the role of the Australian state from protector of society to 'handmaiden of the market' (Bell 1993: 52). In reality, this meant that Australia's relatively small-scale factories no longer had a strategic partner advantaging them in relation to global corporations. Tariffs were cut by 25 per cent in 1974, to an irrelevant 5 per cent by 1993.

This free market strategy led to the rationalization of Australian white goods manufacturing through accelerating acquisitions and mergers, cutting the number of factories from forty in 1954 to four in 2006 under the ownership of the global corporation Electrolux. Widespread factory closures were socially destructive as jobs were lost – not only in the white goods factories, but also in the associated components manufacturers. Furthermore, acquisition by Electrolux only compounded insecurity, since global deregulation has further accelerated acquisitions, mergers and the concentration of production on a world scale; these are processes which exploit space and geographic difference to the limit, eroding the social gains of the past century.

A fundamental distinction between the era of regulated economies and the current phase of free markets is the way corporations now use the spatial fix to impose change at the local level. Restructuring the fridge plant in Orange is a vivid illustration of the leverage that space gives global corporations to force political and workplace change. Restructuring in the Orange factory makes sense when situated within Electrolux's global strategy, which reveals how spatial reorganization and the exploitation of uneven geographic development is the essence of how competition is played out (Harvey 2000: 57).

Within this dynamic a clear geographic pattern emerges, marked by factory closures and relocations. In North America, Electrolux has moved southward to Mexico; in Europe, there is an eastwards drive into Central Europe; in Asia, a movement to China, Thailand and India; and in Australia, a possible shift to China and Thailand. In each case, the magnet for restructuring is the absence of effective democratic unionism and the uneven 'geographical evaluation' of workers (Harvey 2000: 57). These locational shifts can only be slowed to a degree, but never halted, if workers concede to increased work intensification and casualization. Power to impose this unilateral restructuring derives from the highly concentrated structure of the industry in the early twenty-first century.

Electrolux's strategy is determined by the free market policies of national states and the World Trade Organization (WTO), which has generated this geography of production driven by an increasingly concentrated industry, constantly restructured by the spatial competitive dynamic played out through acquisitions, mergers and closures that both create and destroy productive forces. To remain competitive, Electrolux must act aggressively in this process.

The major players are pursuing precisely the same form of competition through leveraging the advantage of uneven geography. Electrolux has pushed the boundaries of how space is used to extend competitive advantage even further. New, previously unimagined, fluid and effective strategies are evolving in these competitive battles in which the Swedish corporation has led the way. Faced with a falling rate of profit in 1996 and 1997, the company engaged in aggressive cost cutting. A target of 6 to 7 per cent was set in terms of an operating margin and around 15 per cent in terms of the rate of return on equity (Electrolux Annual Report 1997: 7). The company lifted its operating margin from 4 per cent in 1997 to 5.2 per cent in 1998, and thereafter reported an operating margin of 6.2 per cent in 1999 and 6.5 per cent in 2000. As Electrolux's 2000 report shows, these results were achieved by a zealous strategy of closures and downsizing that led to stock market values climbing sharply between 1997 and 1999 (Tatge 2000). It is the corporation's new method of achieving these outcomes that further advances how space is being used.

Electrolux's annual reports reveal much about the strategic options available to this powerful global corporation – options based on exploiting geographical difference to the greatest possible degree. During 2002 the company closed cooker factories in Sweden, Italy and Germany, relocating production to Romania; similarly, it has shifted some production capacity from a Spanish refrigeration plant to an established plant in Hungary. The wave of rationalizations continued through 2003. An air-conditioning plant in New Jersey in the United States was closed, resulting in the loss of over 1,300 jobs. In Europe, the company announced plans to close three facilities – a refrigeration plant and a cooking hob factory in Germany and a cooker plant in Norway (Electrolux Annual Report 2002: 17–18). This restructuring continued unabated during 2004. A vacuum cleaner plant in Vastervik, Sweden was closed, resulting in the loss of 600 jobs, and in 2005 a refrigeration plant in Greenville, Michigan was relocated to Mexico, resulting in the loss of 2,700 US jobs.

The geographic pattern of the restructuring demonstrates a movement out of high-waged, unionized Europe, the United States and Australia into union-free, cheap labour zones. In Europe, for example, these waves of divestment, retrenchments and plant closures have meant that Electrolux's employment levels in Northern European nations have declined markedly. Between 1998 and 2002, average employment fell by 30.6 per cent in Denmark (−865 employees), 32.5 per cent in Sweden (−3,163 employees), 38 per cent in Germany (−3,480 employees) and 42.4 per cent in the UK (−1,666 employees). Undoubtedly, Sweden and Germany have experienced the most dramatic absolute decline from the mid-1990s. Taking 1995 as a base year, the average number of employees has declined by 52 per cent in Germany (−6,170 employees) and 56 per cent in Sweden (−8,363 employees) (calculated from Electrolux Annual Reports 1998–2002).

This restructuring is similar to the spatial fixes of the other industry leaders. However, what is innovative in Electrolux is the way the global corporation is shaping the new geography of production through constituting a new form of space competition *within the corporation itself*, a process of 'regime shopping', which signals the next stage of the spatial fix. The dynamic is as follows. When management bargains at a specific factory, they leverage restructuring agreements (intensify labour, downsize and casualize) by threatening closure and relocation to cheap labour zones. This 'whipsawing' strategy is now a cardinal facet of Electrolux's strategy. Significantly, these agreements are of short duration, forcing unions to bargain away conditions each year in the hope that this might influence the company's future 'regime shopping' decisions. The strategy had its genesis in Nike, where the Asian producers contracted to Nike constantly competed for short-term contracts. The crucial difference is that Electrolux has inserted the system within its own companies. An example of whipsawing is Electrolux's recent Italian intervention. The corporation had acquired the leading Italian white goods manufacturer Zanussi. In 1997 the MNC implemented regime shopping, successfully imposing work intensification and casualization after threatening closure and relocation to Romania (Bordogna & Pedersini 1999; EIROnline 2000).

There is a profound power inequality underpinning these changes, which is central to the manufacture of insecurity. Unilateral power is consolidated through the spatial organization of production, in which the production of space and scale erodes union power. Unions accede because there appears to be no alternative, given actual and threatened spatial fixes. This dynamic of accumulation has impacted on the fridge factory in Orange.

Defy in South Africa

Unlike the factory in Orange, the major white goods factories in South Africa are still locally owned. However, this has not prevented local firms from following a similar logic of working space to discipline workers. Local manufacturing became a major casualty when the conglomerates built up under apartheid unbundled and sold off less profitable units. Mining capital shed many of its manufacturing operations and concentrated instead on globalizing its mining operations.

Liberalization started prior to the demise of apartheid. While increasing state human rights repression in the 1980s, the industry policy of the apartheid government began to hesitantly embrace tariff reform and privatization. However, decisive steps could not be taken because of the illegitimacy of the government (Joffe, Maller & Webster 1995; MERG 1993: 214–215). This changed with the opening up of South Africa's economy from the early 1990s onwards. In 1994 South Africa made an offer to the General Agreement on Tariffs and Trade (GATT), and thus committed itself to the liberalization of trade, resulting in manufacturing experiencing severe import pressure.

By the 1990s, Defy was known as TEK Corporation, with factories in Jacobs, East London and Ezakheni. In 1994 it was sold to Malbak, who announced in 1996 that it would embark on an unbundling process. Defy was sold as a separate unit. By 1996 it had a market share of 32 per cent of the South African white goods market, employed a workforce of 2,400, and had an annual turnover of more than R700 million.[2] In January 1997 the Swedish director of Electrolux told the press that the group was 'looking at' Defy. The media speculated about a figure of around R200 million to R300 million.[3] In February 1997, however, it was announced that a local consortium consisting of Defy management and Firstcorp Capital[4] had bought the company for R179 million, 'pipping the world's largest appliance manufacturer, Sweden's Electrolux'. Ross Heron, chief executive officer of Defy, commented that he was 'delighted that the company will remain in [South African] hands because Defy ... is a well-established household brand name'.[5] The new owners immediately invested R15 million in upgrading its Jacobs plant.[6] Referring to ESKOM's electrification programme, Heron commented in late 1997: 'We have an obligation to meet this emergent market. Unit cost must be reduced without compromising quality. As we are now an export nation, Defy is becoming export focused.'[7] Heron refers here to the emerging black middle class, who became a major market for local white goods producers.

When one compares the Australian experience with the Electrolux takeover, Defy's workers in Ezakheni are most probably more secure than had the factory been acquired by the Swedish corporation. Workers at a factory near Johannesburg were less fortunate. One of Electrolux's brands, Kelvinator, was manufactured under licence by Kelvinator SA. This company went bankrupt, and Defy bought its equipment. The company negotiated with employees of the former Kelvinator, and rather than keeping this plant near Johannesburg open, moved the equipment to its Ezakheni operation, where labour was a third cheaper. Here we see how South Africa's internal uneven geography is again used by the corporation (Bezuidenhout 2006).

While white goods are still manufactured at a number of factories in South Africa, the volumes produced are small relative to the emerging global economies of scale. There is some investment from outside the region, when Whirlpool bought a loss-making factory, but quite unlike the Australian trend, corporations generally remain in the hands of South African entrepreneurs. Since the 1990s, several firms and factories closed down, and there is a growing concentration in the industry. The South African government has reduced import tariffs, and is committed to reducing them further. Local firms generally produce for local markets, and imports have increased at a rapid pace. Much of the country's downstream manufacturing industry was wiped out by rapid tariff liberalization. Defy is holding out against all odds and it is not clear whether or not the firm will remain in South African hands (Roberts 2006).

LG in Korea

Whereas in Australia Email was taken over by the multinational Electrolux and the Orange plant dramatically downsized, and in South Africa rationalization was driven by local corporations locked into a local market, in Korea LG successfully globalized and became a top ten player in the global white goods market.

In the late 1980s, the LG's global expansion drive gathered pace. In 1989 a sales subsidiary and a joint production subsidiary were established in Thailand and the following year an Ireland-based design technology centre was created. In 1993 a subsidiary was established in Huizhou in China. In 1995 the Lucky Goldstar name was changed to LG Electronics and the re-branded company bought the US-based Zenith Corporation.

In 1997 a production plant was opened in India and a joint venture with Philips LCD was established. In 2000 the corporation launched the world's first internet-enabled refrigerator, and their global sales of refrigerators led the entire market. In 2002, under the LG Holding Company system, the company spun off to LG Electronics (LGE) and LG Electronics Investment (LGEI).

LG's white goods manufacturing operations currently fall under LG Electronics, Inc., a subsidiary of what is termed in Korea a chaebol or conglomerate. According to LGE's website, 'the company is a global force in electronics, information and communications products with 2004 annual sales of US$38 billion (consolidated).' It employs 'more than 66,000 employees working in 76 subsidiaries in 39 countries around the world.'[8]

LGE has made it clear that the firm wants to increase its global market share. Its vision is to be 'Global Top 3 By 2010'. The company's website explains:

> LG Electronics sets its mid-term and long-term vision anew to rank among the top 3 electronics, information, and telecommunication firms in the world by 2010 ... As such, we embrace the philosophy of 'Great Company, Great People,' whereby only great people can create a great company, and pursue two growth strategies involving 'fast innovation' and 'fast growth.' Likewise, we seek to secure three core capabilities: product leadership, market leadership, and people-centered leadership.[9]

LG had truly become a global corporation, and with all the talk of 'great people' and 'people-centred leadership', the company employs strategies not unlike Electrolux, where workers in different countries are played off against each other. Apart from its operation in Changwon, the company now manufactures refrigerators in India, Indonesia, China, Vietnam and Mexico. It also has production facilities which manufacture other white goods product lines in Turkey, Kazakhstan and Brazil. It has established an additional two plants in China (LG Electronics 2005: 14). In Korea itself, labour law reforms by the government enabled the firm to restructure its workforce by increasing the number of irregular workers, as shown in chapter 3.

Perils of the Dutch Disease

A key factor in explaining the difference in the strength of manufacturing in the three countries lies in what has become known as the Dutch Disease (Auty 1995). Exponents of the Dutch Disease argue that the potential for natural resources, and in particular minerals, to contribute to accelerated economic growth can be seen as a curse. Indeed, research comparing the experiences of mineral-rich countries in Africa and Latin America with the resource-poor but economically successful Asian 'tigers' provides strong statistical evidence for an inverse relationship between extensive mineral endowments and strong economic growth (Auty 2004). A comparison of the manufacturing success of resource-poor Korea with that of mineral-rich South Africa and Australia is a case in point.

As can be seen in table 2.2, South Korean manufacturing firms such as Samsung, Hyundai and LG are all listed in the top FT 500. None of

Table 2.2 FT Top 500 corporations measured as market value (US$), 2006: Australia, South Africa and South Korea

	Australia			South Africa			South Korea		
	Corporation	Market value $m	Sector	Corporation	Market value $m	Sector	Corporation	Market value $m	Sector
	BHP Billiton (32)[13]	116,692.40	Mining	Sasol (273)	25,746.80	Oil and gas production	Samsung Electronics (35)	107,197.10	Technology hardware and equipment
	Rio Tinto (61)[14]	79,423.40	Mining	Anglo Platinum (387)	19,836.70	Mining	Kookmin Bank (236)	29,045.60	Banks
	Commonwealth Bank of Australia (159)	41,628.90	Banks	Standard Bank (407)	18,611.20	Banks	Korea Electric Power (260)	26,939.70	Electricity
	ANZ Banking (193)	34,562.10	Banks	Firstrand (418)	18,217.40	Banks	POSCO (326)	22,477.50	Industrial metals
	Telstra (198)	34,248.80	Fixed-line telecoms	MTN Group (469)	16,622.60	Mobile telecoms	Hyundai Motor (389)	19,763.10	Automobiles and parts
	Westpac Banking (220)	31,105.10	Banks				SK Telecom (476)	16,300.40	Mobile telecoms

(Continued)

Table 2.2 (Continued)

	Australia			South Africa			South Korea		
	Corporation	Market value $m	Sector	Corporation	Market value $m	Sector	Corporation	Market value $m	Sector
	Woodside Petroleum (344)	21,619.50	Oil and gas production				LG Philips LCD (483)	16,129.60	Electronic and electrical equipment
	Westfield Group (349)	21,340.30	Real estate				Shinhan Financial (488)	16,081.40	Banks
							Woori Finance Holdings (494)	16,010.00	Banks

Ranking on the FT index in brackets

Source: Compiled from *Financial Times* Global Top 500 special issue: 'The world's top 500 companies in 2005', June 2006

Australia or South Africa's manufacturing corporations make the list. In contrast to Korea, it is firms involved in mining and finance which dominate, followed by service companies. In addition to the dominance of mining and finance capital in South Africa and in Australia, one should note that a number of former South African firms also feature, but since they were allowed by the post-apartheid government to move their primary listings to the economic core, they are no longer considered to be South African. This includes the Anglo American Corporation,[10] a mining holding firm, and SAB Miller,[11] the result of a merger between South African Breweries and the Miller brewery of the US. Both Anglo American and SAB Miller are now listed in the United Kingdom. Another firm with South African origins made the list, namely Richemont,[12] the tobacco and personal goods empire of the late Afrikaner capitalist Anton Rupert. Its primary listing is in Switzerland.

The fact that several of South Africa's largest mining and financial houses relocated their primary listings at the advent of democracy to London, and in some cases New York, is significant for local configurations of class power. Also, Australia's two top corporations are technically transnational corporations, since they are also listed in the United Kingdom. This underscores the persistent imperial nature of the dominant mining and finance class factions in both countries. Not surprisingly, when one considers the differences between the trade structures of the three countries, these differences become even clearer, as seen in table 2.3.

In Australia and South Africa – countries rich in mineral resources and where mining and finance capital is dominant – there is little interest in nurturing and protecting local manufacturing operations (Fine & Rustomjee 1996). In contrast, South Korea succeeded in building up

Table 2.3 Indicators of trade structure, 2004

	Imports of goods and services (% of GDP)	Exports of goods and services (% of GDP)	Primary exports (% of merchandise exports)	Manufactured exports (% of merchandise exports)
Australia	21	18	58	25
South Korea	40	44	8	92
South Africa	27	27	42	58

Source: UNDP Human Development Report 2006

an impressive base of manufacturing firms that have become world players. The social disruption caused by the Korean War allowed a capitalist class to be constructed anew. As we argued at the beginning of this chapter, another crucial factor in the global success of Korean corporations such as LG is the counter-market interventions of the Korean state and its militaristic-like action against organized labour to maintain the global cost advantage of Korean labour against countries such as South Africa and Australia that have democratic trade unions.

Conclusion: The Global Pursuit of Shareholder Value

In his provocative analysis of the global corporation, Bakan (2004) argues that corporations are 'morally blind', and act with a 'reckless disregard for consequences' that result from the organization's flawed institutional character. He shows how this character is determined by state intervention through corporate law which requires a singular focus on shareholder value. Bakan concludes that the corporation, 'like the psychopathic personality it resembles, is programmed to exploit others for profit. That is its only legitimate mandate' (Bakan 2004: 22, 59, 69). A key trait of the psychopath is their inability to concern themselves with their victims. Corporations evidence a similar detachment from the victims of restructuring. Our analysis of corporate behaviour in this chapter attests to this argument.

In Australia, the white goods factory in Orange after the Second World War became a site of stability for working people. From 2001 onward, this fulcrum of security wanes as the firm is acquired by Electrolux. Driven by market ideology based on the aggressive global pursuit of shareholder value, the factory is restructured into an insignificant assembly operation, with no guarantee that it will remain in Orange. Despite the significant job loss which destroyed the basis of life for these workers in Orange and in every other factory across the globe where the company has instituted a similar process, Electrolux proclaims that it is a socially responsible company that recognizes basic trade union rights. Certainly, the global company has made all the right moves. It claims to have developed corporate policies that are consistent with the values of Swedish social democracy. It is a member of the United Nations Global Compact, which states, 'Businesses should uphold the freedom of association and the effective recognition of the right to collective bargaining' (Section Three). The company has also formulated a Workplace Code of Conduct, which was adopted by Group Management in early 2002. The code defines minimum acceptable work standards and is based on internationally recognized treaties and agreements, such as the core conventions of the International Labour Organization and the OECD Guidelines for Multinational Enterprises. The code includes a commitment to freedom of association and the company states that its suppliers are required to comply with the code. But, as is clear from Orange and Electrolux's restructuring across the globe, these noble principles are empty rhetoric, as the company works space relentlessly in its pursuit of shareholder value. In this the Swedish corporation is no different to any other global corporation.

Highlighting this contradiction foregrounds a critical issue that is the subject of Part Three. Global corporations such as Electrolux refuse to negotiate restructuring and their spatial fixes with unions. On the European Works Council they are ready to brief delegates on their plans, but assert that restructuring is off the table as it is a managerial prerogative. The worldwide crisis of declining union densities is not unrelated to this predicament. Workers we interviewed stated that unions had become irrelevant because they lacked the capacity to intervene where it really counted. The possibility of developing strategies to engage restructuring is considered in Part Three.

As in Australia, South African manufacturing is also under severe pressure from imports. Unlike Australia, South Africa's major white goods manufacturer remains in local hands. Defy has developed as a strong local white goods brand. The apartheid state, responding to the threat of independent black unions, offered subsidies to firms who located their manufacturing operations in rural areas, controlled by pseudo-independent tribal regimes, and Defy opened a major manufacturing operation in Ezakheni. When the apartheid regime crumbled and the economy opened up to competition, several white goods manufacturers went bankrupt, with Defy emerging as the market leader. However, the industry is squeezed by an increase in imports – mainly from China and South Korea.

In South Korea, the state's involvement in the economy is more hands-on. LG is one of the major chaebols that emerged out of the military regime's industrialization programme. The industrial complex in Changwon was created by the policies of this regime. The firm has grown from a local firm to become one of the leading global white goods manufacturers. LG is committed to maintaining its operations in Changwon, but has recently started to move component manufacturing to lower-wage countries such as China. Its corporate strategy involves managing a complex web of suppliers, as well as labour hire firms. This creates a sense of insecurity among LG employees.

We turn now to an analysis of how the global restructuring of the white goods industry is transforming the three workplaces in Changwon, Ezakheni and Orange.

3

The Return of Market Despotism

Manufacturing refrigerators is relatively straightforward. You build a metal cabinet, insulate it, and fit shelves, a door, compressor and condenser. The compressor is the most expensive component. In recent decades, environmental concerns have forced firms to create CFC-free refrigerators, and frost-free fridges have become fashionable. In terms of design, the future and the past play a role. There is a move to futuristic stainless steel refrigerators for the upper end of the market – undermining the idea that all refrigerators are necessarily *white* goods. Some retro designs attempt to recover the nostalgia of previous decades, with some sixties designs in shades of pastels appearing in the pages of kitchen design catalogues. While there are high profile innovations, such as LG's refrigerators that are integrated with television and internet technology, these extremely expensive models remain essentially marketing tools. There are very few households which have refrigerators that can *actually* sense what is on the shelves and dial onto the internet to order new produce from shops, who then deliver the goods to hi-tech homes.

While the technical business of manufacturing refrigerators is rather simple, the industry itself is more complex. As the previous chapter showed, the firms involved in the manufacturing of white goods in Ezahkeni, Changwon and Orange have all undergone restructuring. There has been a spatial reorganization of the industry, with production increasingly being concentrated in mega factories that are close to emerging markets such as China, Brazil and Turkey. These production platforms are then used to export to more traditional markets (Nichols & Cam 2005).

Global industry restructuring is impacting on the lives of workers in profound ways. This chapter is about how workers experience the changing world of work where they assemble refrigerators and freezers. The LG factory in Changwon is one of the most productive in the world. In 2005 it produced 2.8 million refrigerators, mainly for the export market. The Defy factory in Ezakheni is mostly locked into production for the South and Southern African market. Electrolux's factory in Orange also manufactures largely for the Australian market. Most of the workers interviewed for our study worked on these assembly lines.[1]

The 'double movement' implies a constant tension between treating labour as a commodity and struggles to de-commoditize labour. In order to ground these dynamics in the workplace, we have to take a short detour in order to consider theories of the labour process. The tension between treating labour as a pure commodity and the need to protect it from such excesses is reflected in regimes of control in factories, retail outlets, call-centres, mines and farms. Burawoy (1979, 1985) broadly distinguishes between *despotic* and *hegemonic* regimes of control. Despotic forms of control draw on coercive power, both physical and economic. The effect of these forms of coercion is to create feelings of negative self-worth among workers, since it is an attempt to treat labour as a pure commodity. Hegemonic forms of control are designed to 'manufacture consent' – or at least compliance. Early capitalism in the industrialized countries – during the First Great Transformation – essentially constructed regimes of control around *market despotism*. The whip of the market was used to discipline workers. If they did not perform, they were dismissed. Since workers were treated like commodities – as objects – they lacked voice in the workplace and hence there was no regulation of conditions at work.

However, as Polanyi points out, such harsh forms of labour commoditization led to a counter-movement. Society responded by making certain demands on employers and the state. Certain conditions, including warfare, allowed the countries of the North to strike a historical compromise between capital and labour in the creation of welfare states. This resulted in a shift to *hegemonic* regimes of control which are based on the regulation of working hours, the setting of minimum wages, putting in place health and safety standards, and mechanisms for trade unions to organize and bargain collectively over wages; in short, various ways in which labour is de-commoditized and made less insecure. Central to this shift is the emergence of forms of counter-power to the power of management.

As the Northern class compromise came under pressure in the 1970s and 1980s – the Second Great Transformation – so did hegemonic

Table 3.1 Shifts in regimes of control in the North

First Great Transformation	Counter-movement	Second Great Transformation	Counter-movement
Rapid marketization and commoditization. Market despotism in the workplace	Emergence of workplace hegemony and construction of a welfare state	Rapid liberalization and shift to hegemonic despotism	Embryonic global counter-movement in the post-Seattle period – World Social Forum, new global unionism

regimes of control. Burawoy argues that these made way for what he calls *hegemonic despotism*. This implies that the institutions of collective bargaining are now utilized to concession bargain, where workers agree to the re-commoditization of their labour under the threat of factory closures or lay-offs. The ideology of globalization legitimizes this (see table 3.1).

Nichols et al. (2004) have criticized Burawoy's framework for not being sensitive to issues of the employment contract. In their comparative study of white goods factories they demonstrate how the segmentation of labour markets into a core of permanently employed and a periphery of insecure contract workers has emerged as central to the system of control in the workplace. The implication of this observation is that it is possible to use a combination of hegemonic and despotic regimes of control in one factory. Regimes of control are characterized by boundaries that exclude workers on the basis of identity (of citizenship, race, gender, etc.), or on the basis of territory (of country, industry, city, etc.). For instance, white miners at Lambing Flat in Australia attacked their Chinese competitors in 1861. This resistance to competition from non-Europeans led to a policy of only allowing European immigration. This policy continued until the 1970s. As a consequence, Australia's welfare state was based on the exclusion of non-Europeans from the country. Similarly in South Africa, after the Rand Revolt of 1922, when white miners protested against the competition mine owners created by employing black workers at lower wages, black workers were formally excluded from the class compromise between capital and white labour until 1979.

The Northern class compromise did not involve their colonies. In the colonies the possibility of establishing hegemonic forms of control was constrained by coercive labour practices. Here the workplace regime was often based on what Burawoy calls *colonial despotism*. Because colonialism only partially penetrates society and only partially proletarianizes its subjects, options outside wage labour are still available to disgruntled workers. Hence, coercive measures, including compounds and restrictive contracts of employment, are used as a form of control. The importance of 'race' in the occupational hierarchy and supervision should not be underestimated in the construction of this regime, as vividly illustrated in the South African case (Webster 1985; Von Holdt 2003).

Counter-movements in the colonial world – the Global South – often dovetailed with struggles for national liberation. Once this is achieved, post-colonies are faced with the dual problems of demands for a change in the workplace regime and scarcity. Possibilities for establishing hegemonic regimes of control are therefore constrained. In Africa, post-colonial states often accord workers certain rights and guarantees, but the majority of the population are excluded from this as a rule, since they work in the informal sector, or are unemployed. Labour movements are often integrated into the post-colonial state, in what could be called a form of state corporatism. When this is challenged by neoliberal globalization, usually in the form of an International Monetary Fund (IMF) imposed 'structural adjustment programme', the labour movement is one of the first to come under attack, and often ends up in opposition to former comrades from the liberation struggle. State corporatism comes under pressure in the name of 'labour market flexibility' and the assumption is now that those in formal employment are part of a labour elite.

As illustrated in table 3.2, countries in the South have followed a historical trajectory that differed markedly from the First Great Transformation of Northern industrialized nations. The history of the South is marked by the colonial experience of political and economic subordination to the needs of the Northern capitalist economies. As Barchiesi (2006) argues, at the core of the welfare state of advanced capitalist society was a link between wage labour and social citizenship. The 'social question' was solved and workers' demands were met by the introduction of the welfare state that began a process of redistribution through state transfers. However, in the South, he suggests, colonialism could not deal with the 'social question'. These countries lacked the preconditions for a successful resolution of the social question, a political coalition made of strong unions, well-organized employers and a government that considers industrial citizens its core constituency. In Polanyian terms, they skipped a stage. These societies never secured a welfare state, high-waged employment and social citizenship as their

Table 3.2 Shifts in regimes of control in the South

First Great Transformation	Counter-movement	Second Great Transformation	Counter-movement
Colonial conquest and land dispossession. Colonial despotism in the workplace	National liberation movement. Leads to political independence and state corporatism	Structural adjustment, market despotism	Embryonic global counter-movement in the post-Seattle period – World Social Forum, new global unionism

own democratic transition occurred at the very moment of the Second Great Transformation. Political liberation was secured within the global environment of market-driven politics and restructuring of work and society.

While South Africa, South Korea and Australia have divergent histories, and different regimes of control, shifts over time are leading to a convergence around a profound sense of insecurity. In each of the three countries workers have managed to gain some protection as a result of their collective struggles. Australia was the first nation in the late nineteenth century to introduce an 8-hour working day and later a universal welfare state. In Korea, the labour movement played a key part in the democratization movement culminating in the fall of the military government in 1987 and the extension of rights to organized labour. In South Africa, the labour movement was central to the demise of apartheid, and workers chipped away at the pillars of apartheid in their workplaces even before the system was formally abandoned in 1994. It was able to establish new labour laws that formalized many of these gains, including entrenching the right to strike in the constitution.

But in each of the three countries these gains came under severe pressure in the 1990s. In Australia, the Labour government dismantled the industrial policies on which its hegemonic labour regime rested. Unions entered into concession bargaining, which was sold to union members as 'strategic unionism'. It was left to the Conservative government of John Howard to dismantle the labour laws to allow for a return to market despotism in 2005. Both South Africa and South Korea have undergone double transitions, where political democratization coincided with economic liberalization.

In all three factories there has been an increase of casualization through the introduction of various forms of non-standard contracts of employment. Also, work has been intensified, and workers toil under the constant threat of a spatial fix, or relocation. Yet, this impacts in divergent ways in these different locales.

In this chapter, we attempt to make sense of how, in the age of insecurity, this restructuring *manufactures insecurity* by creating flexible worlds of work. In particular, we show how this undermines the power of labour, both associational and structural bargaining power. We begin with LG in Changwon, the most extreme version of a flexible workplace.

Changwon: Life is Good ...

The popular uprisings in the late-1980s in South Korea profoundly shaped the workplace regime at the LG factory in Changwon. Prior to 1989, the factory operated along despotic lines. However, two militant strikes – one in 1987, and another in 1989 – forced LG's management to rethink their approach to labour relations and production. The 1989 strike is sometimes likened to a war situation, and the riot police were called in to contain matters. LG's current CEO, Kim Sang-su (or SS Kim, as he is popularly referred to) was a white-collar employee at this plant at the time and was central to devising the new approach.

When workers arrived at the factory gate after the 1989 strike, they were surprised to find the entire management team waiting their arrival. As they entered the premises, management greeted them with a bow – a sign of respect in Korean culture. After lunch, management cleaned their canteen. At first they were sceptical of these symbolical gestures, but when management sustained these actions over months, they accepted the fact that there was a new approach at LG. Also, management initiated a new approach to work organization. Workers were sent to the Matsushita plant in Japan to study this successful factory's approach to production. Following the Japanese orientation to manufacturing, workers were given relatively high wages, very good company benefits, as well as job security in return for their commitment to high productivity and cooperation with management. Where other chaebol managers were profoundly anti-union, LG pre-empted unionization by democratic unions by introducing a company union.

A number of innovations were pioneered at the factory. The assembly line was cut from 120 metres to 60 metres, and modular production was introduced in certain key areas. LG was also a key driver and developer of the Six-Sigma approach to production, a strategy pioneered by

General Electric which has become popular globally. This technique uses statistical methods for quality control to improve all processes, including human resources issues such as absenteeism rates. More recently, the company has introduced a strategy called TDR – Tear Down and Redesign – developed by SS Kim. Workers are taken from their jobs and told to improve work processes in TDR teams – meaning that that they are encouraged to completely redesign certain targeted parts of the production process. Winning TDR teams receive prizes, and members are often rewarded by going on study tours to LG's plants in other parts of the world.

Apart from new approaches to production and the involvement of workers in these decisions, LG takes care to establish the company's hegemony symbolically by presenting workers with paraphernalia such as watches for long service. While not greeting workers at the gate any longer, management has seen to it that the union plays a key part in selling the company line to employees. The persona of SS Kim looms large – he is seen as coming from the workers, and not the LG chaebol royalty, who own the conglomerate. The company is driven by the ambition to become the top global manufacturer of digital appliances by 2007. The chairman of the local branch of the company union refuses to talk about 'labour relations', insisting that LG has a 'labour-management relationship', based on 'mutual trust', a 'win-win relationship'. He proudly referred to the competition between LG and Electrolux:

> We know Electrolux are feeling threatened. They gave stickers to their employees to say that LG is following them to attach to their rear view mirrors.

In recent times, and especially after the financial crisis of 1997, the factory has come under considerable pressure due to rising costs of production and competition. A number of strategies other than Six-Sigma and TDR have been employed to keep costs down. While management has signalled its commitment to keeping the factory in Korea – unlike Samsung, who have moved most of their operations to China – LG uses a number of more despotic strategies to supplement the win-win relationship in the factory.

First, it should be noted that since the early 1990s, 30 per cent of LG's workforce are now irregular workers. These workers were never part of the core workforce who receive good benefits and who participate in production decisions. The company has gradually increased the proportion of irregular workers, especially after the financial crisis.

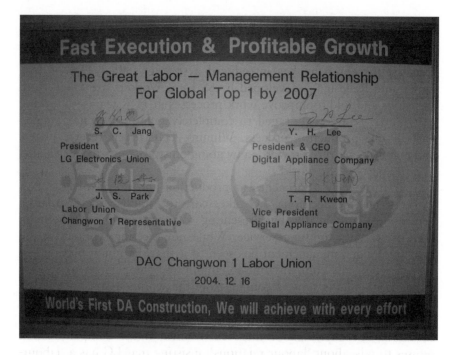

Figure 3.1 Agreement between management and the LG company union in Changwon to achieve the goal of becoming the global leader by 2007 (photograph: Andries Bezuidenhout)

In the late 1980s, when the firm initiated its restructuring programme, the factory employed more than 7,000 staff. By early 2000 it had reduced the workforce by half, which included 1,000 irregular workers, who mostly work for internal contractors. Second, LG has squeezed its suppliers through a strategy known as CR, cost reduction. Some suppliers have relocated to China in order to contain their costs. The firm's position in the commodity chain accords it the power to actually demand cuts in costs from its suppliers.

These two aspects of the production regime – the use of irregular employment, as well as a complex web of supply chains – intersect to create an extreme form of labour market segmentation. Different types of employment contracts are embedded in the industrial structure of LG's operations – a structure that is very typical of the way in which most of the major manufacturers in Korea operate. Beverly Silver (2003) calls this a 'lean and dual' approach to production, in contrast to the

Western adaptation of the Asian model as 'lean and mean'. A core of highly incorporated employees forms the basis of the production system. They are members of a company union that identifies closely with management objectives and culture. These workers are paid relatively high wages, but are supplemented by a secondary labour market of 'irregular' workers.

Comments from workers in this first tier of the labour market reflect the extent to which management has succeeded in incorporating them into company ideology and practice. 'A win-win relationship has been made', argued one. 'Management has shown the transparency and trust to workers. The more cooperative relationship has been achieved, I think.'[2] Another said: 'The relationship is good and much cooperative. We have maintained "smooth" relationship. I believe that the management has been trying to show transparency and respect for workers ... labour and management have been trying to improve competitiveness together.'[3] Another commented: 'The relationship has been consistently cooperative.'[4] Nevertheless, while these comments from workers tend to reflect the official company position, one of the workers interviewed was somewhat critical. He said:

> On the surface, the relationship has been more cooperative. But I think the management has overwhelmed employees like me. The management has argued that labour and management are equal partners. I don't believe it. Employees have had to unilaterally accept the demands of restructuring and retrenchment by the management. Many regular workers have been replaced with irregular workers of internal/external outsourced companies.[5]

Indeed, the tendency has been to replace regular workers with irregular ones, who work for what is called in rather Orwellian fashion 'cooperation companies'. These are LG's internal contractors, who employ another 1,000 people. Examples of these firms were mentioned in our interviews, with names such as Geumjin Industry, Hongseon Industry and Shinil Industry cropping up. However, workers generally saw these firms as quite unstable, as explained by one of their employees:

> The name of my company has changed many times. It was originally the Seungwon Industry. Now, it is the Geumjin Industry. The LG company forced the internal outsourced companies to merge, sometimes to split. Among the internal outsourced companies within the LG company, the procedure of merger and split has been repeated many times. This means my relationship with management is not stable. It is very precarious beyond my expectation.[6]

Another explained how this type of employment reinforces feelings of insecurity:

> Frankly speaking, no regular worker of the LG wants to be a worker of the internal outsourced companies. It means that we become a 'second class' of employees of the LG company. Five years ago, I was a regular worker with more employment security. Also, the management was big-sized company with the strong socio-economic power. But, now, I am working for small-sized company with weak socio-economic power, which is actually controlled by the mother company, the LG.[7]

Another agreed:

> Actually, my company is controlled by the mother company, LG. It can be said that the real employer for me is not the internal outsourced company, but the LG itself. I don't understand [why] this kind of illegal situation is happening, as the government is looking on with folded arms.[8]

But the story does not end here. The segmentation of a workforce into a core and a periphery is also replicated by LG's suppliers. The assembly plant at the LG complex is what the term suggests – an assembly operation. Major parts of the production process are located at the premises of its suppliers, who are required to adhere to certain standards and processes laid down by LG's management. The regular workers of these suppliers generally do not belong to trade unions – enterprise unions or otherwise. According to a number of our respondents, LG is seen to be quite strict about this. If the regular workers at one of its suppliers openly join trade unions, those suppliers will lose their contracts. The workers of one of the LG suppliers we interviewed did not want to affiliate to any of the union federations because of this concern. Indeed, theirs was the only supplier with a union that organized its regular workers. Our interviewees in this segment of the labour market, even though they were officially classified as 'regular workers', did not talk about 'win-win' relationships with their management. When asked about their relationship with management, they described it as such: 'It has been consistently confrontational'; 'I am not interested in management'; 'The relationship has been consistently strained'.[9] One of these workers elaborated on the underlying reasons for this confrontational relationship:

> In my assembly line, there are many regular workers. For this reason, the management does not like my line. The management wants to move the line to China or outsource it to different companies to reduce the number of regular workers. We have this kind of tension with management.[10]

As in the case of LG, roughly half the workforce of this supplier was employed through internal contractors such as Star Industry, Human Tech and Wal Tech, and hence these workers were classified as irregular workers. One of the workers we interviewed in this category had formerly worked as a regular employee for LG, but was retrenched.[11] As in the case of LG's own irregular workers, a deep sense of insecurity prevailed. Indeed, management is seen as aloof and uninterested in irregular workers altogether. A worker mentioned that management paid 'no attention or personal interest to workers like me'.[12] Another said: 'Management [pays] no attention to its irregular workers.'[13] Another asserted an identity as a 'co-worker', a theme that is quite common in attempts of Korean irregular workers to counter prevailing notions that they are somehow not 'workers': 'They haven't paid enough attention to us as co-workers. I think it has taken long time for me to be accustomed to the management's atmosphere and characteristics.'[14] A sense of being 'ignored' and 'frustrated' prevailed: 'Usually, management has ignored and frustrated us from my beginning days for the company.'[15]

These suppliers outsource a part of their production process to even smaller suppliers. The last layer of suppliers – in effect these are suppliers to suppliers – are at the bottom of the food chain, or rather the production chain. They employ very few workers – often as little as one or two. They tend to operate from basements located under houses in the suburbs of Changwon. They generally do not exclusively supply parts or components to one firm, but have multiple clients.

Here we see how labour market segmentation is embedded in the industrial structure of production in Changwon. We have identified five layers of employees, which could be described as a cascading process of outsourcing. To summarize, the links in the chain are: (1) LG's regular workers, (2) LG's irregular workers, (3) regular workers of various suppliers, (4) irregular workers of various suppliers, and (5) workers for suppliers of suppliers in small basement operations. This is illustrated in figure 3.2.

In Changwon, the overwhelming majority of our interviewees felt that they had to work harder than five years ago. It seems as if the intensification of work was more prominent among irregular workers in LG and the suppliers. Nevertheless, LG's core workforce was very aware of competition in the industry:

> I feel that I have had to work harder. The goal of my company is to maximize profits, which is the top priority of the company. The management

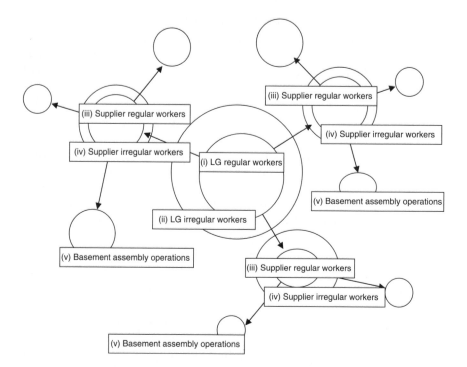

Figure 3.2 Labour market segmentation and industrial structure at LG

has implemented the processes of innovation of production and organization. The competition between companies as well as workers has been increased.[16]

It is interesting to note here that competition between companies also means competition between workers. Another worker mentioned the threat of moving the plant to other countries:

> We are working harder. The competitiveness of my company has been weakened. Also, some plants were moved to overseas countries. The management has enforced employees to improve productivity. I feel mental and physical fatigue.[17]

In contrast to the official line that management and workers are equal partners, an interviewee talked about the 'top-down' nature of work intensification:

I am working harder. Also, the management adopted top-down style in production line. This means the management set up the target amounts for production without consultation with employees.[18]

An interviewee who worked in a supervisory position talked about the mental burden brought about by demanding that workers work harder:

As I am a chief of my section, I have to manage and lead workers. For this reason, I feel mental burden rather than physical one. Generally speaking, our workers have had to work harder, because of the intensified competition in white goods market.[19]

Similar perspectives emerged from interviews with workers employed by LG's internal contractors. One mentioned that he has 'become old' from the 'physical burden'.[20] Another mentioned the threat of relocation:

I have been working harder. The managerial situation of my company has been deteriorated. The competition in the market has been seriously widened and deepened. The mother company, LG, has been trying to reduce the products of white goods and move the assembly lines to China.[21]

LG has also been squeezing its suppliers. Workers interviewed in one of the supply firms linked the intensification of work to their being inserted into the LG supply chain: 'The LG company has demanded our company to increase the products to meet its demand. As a result, job intensity has gradually increased.'[22] Another mentioned 'neoliberalism' and 'globalization', and linked this to the growth of irregular employment:

It is true for me to work harder. I think it results from neoliberalism and globalization. The amount of products has rapidly increased. Quality management has been very strict. The retrenchment of regular workforce has been made. The jobs of regular workers have been replaced by irregular ones.[23]

The above worker was a member of a trade union, which often interacted with the local structures of the Korean Confederation of Trade Unions (KCTU), but did not affiliate formally to the federation for fear of victimization. The KCTU represents the new democratic union tradition that emerged openly after the 1987 strikes, whereas LG's company union is affiliated to the Federation of Korean Trade Unions (FKTU) established after the Second World War by the United States and Korean establishment.

As discussed in chapter 8, it was designed as a mechanism of control over the militant labour movement at the time rather than as a voice of workers.

LG workers also mentioned the recent introduction of a five-day working week in Korea as contributing to work intensification:

> My company introduced 5-day work week system a year ago. But we cannot have a rest during weekend [because we have to work overtime]. The management has demanded workers to produce the same amount of units in five days, which was produced in six days before the introduction of 5-day work week. Many regular workers have complained about increased labour intensity, criticizing the introduction of 5-day work week.[24]

Some interviewees were able to give us a real sense of the extent to which work has intensified. An employee of one of LG's supplier's internal contractors said: 'At that time when I came to [this company], 8 months ago, I made an assembly of 40 units an hour. But yesterday, I did 80 to 85 units an hour.'[25] Another said: 'At that time when I came to the [company], 3 years ago, I made decoration assembly of 300 to 350 units in two hours. But yesterday, I did 400 to 450 units in two hours.'[26]

The LG plant is one of the leading manufacturers of refrigerators in the world, but it is also one of the leading manufacturers of insecurity. This leads to employees working longer hours, harder and faster in the hope that the factory does not close and relocate elsewhere. In Korea we have seen a shift from what could be described as state despotism to an incomplete form of hegemony characterized by a 'lean and dual' approach to the workplace. This approach is characterized by core workers who are incorporated into a company union and company welfare side by side with a growing number of irregular workers who are excluded from this. In the mid-1990s, LG aggressively globalized its production platform, and even core workers feel threatened by the possibility of production being relocated to China or elsewhere. The Asian financial crisis in 1997, and the IMF intervention demanding the flexibilization of the labour market, added to this as the number of irregular workers was increased (Yun 2007). Chang (2002) calls this authoritarian flexibility; we call it market despotism.

In this restructuring, LG appears to have set something of a global benchmark of work intensification. In fact, the LG factory in Changwon has become a role model of sorts that other firms seek to emulate. When Electrolux acquired the Orange fridge factory in 2001, they sought to emulate LG. In an interview in 2003, Electrolux's Director of Australian Manufacturing, Leon Andrewartha, told us LG is 'ripping the heart out' of the refrigerator factory in Orange. He said LG posed a real competitive threat. They were now 'light years' in front. He went on to say:

> We are basically now using very similar plant equipment. But they run at 10 products per person per day. We're running at 2.2–2.3. [It is] because of work tempo, they've got the volume; they're running a fridge every 12 seconds ... They build their factories from the ground up so that they don't have to deal with the paradigms of the past. But the plant and equipment in it are exactly the same. So what it comes back to is the way we use it.[27]

In this we witness the creation of global structures, methods of production and benchmarks that create deep insecurity and dehumanize workers, reducing all to a commodity status, a thing to be pushed to the outer limits, just as a machine is purchased for the speed it can spew out product. In Part Three we ask can unions engage these issues in new ways? Can alternative global benchmarks be established, ones that humanize work experience instead of degrading it?

Orange: Nothing Sucks Like Electrolux ...

From its establishment in 1946 until its acquisition by Electrolux in 2001, the fridge factory in Orange created opportunity and personal security for working men and women who had moved to the town in search of a new life. A sense of place is strong in Orange. Sandy, an assembly line worker reflected, 'I love Orange with a passion'.[28] Workers crossed the Blue Mountains of New South Wales in search of something different, turning their backs on the pace of city life for rural rhythms, the beauty of the landscape, gently rolling hills and valleys, rich basalt soils, fertile orchards. All interviewed indicated that the choice to move to Orange was a lifestyle choice. Some bought smallholdings and cultivated their land during leisure hours. Without the work that the factory provided, these lifestyle options would remain a fantasy.

The history of the 65-year-old factory is intertwined with this sense of place. Without the factory, the town would never have attained its economic vitality, for 1,800 jobs in a country town produces a significant multiplier effect. Sixty years ago, Orange workers determined the geography of white goods production in Australia when they united in their opposition to the closure of the munitions factory the government had established during the Second World War. When closure was announced, the town was 'seething with anger'. Well-attended, vociferous meetings resolved, 'If the government can't run the factory then we workers will organize and run it ourselves'. The threat led to a sit-in strike (Edwards 1996: 1, 7). The state government conceded and the fridge factory was established.

Workers' feelings of security are integral to their sense of place. The plant, located near the town centre, was the largest manufacturing facility in New South Wales. Over the decades, threats of closure were absent from the discourse and would have seemed absurd, given the corporation's Australian market dominance. This sense of security was reinforced by the union rights conceded by the Australian industrial relations system. Local unions were strong and were unafraid to strike. A union delegate reflected on the culture of solidarity that shaped working life:

> In the past they couldn't put it over you. They knew you were not going to back down. In the past if they tried to push us we would have all been out on the grass for weeks. We used to organize a barbeque out in the front – sort of in their faces, you know.[29]

Edwards (1996) uncovers the historical genesis of this confident, assertive culture. Over the decades unions fought for 'reasonable working conditions', winning a 38-hour week, holiday leave loading, pensions and equal pay for women (Edwards 1996: 61). As a result of these gains, workers regarded the factory as 'a reasonably humane place to work'.[30] Two unions of roughly equal membership represent workers in the plant. The Australian Workers Union (AWU) is a more traditional union and has often been willing to compromise over restructuring. In contrast, the Australian Manufacturing Workers' Union (AMWU) is a more militant and innovative union, as is discussed in chapter 7. The AMWU used its strength at the factory to pioneer agreements that were used to set standards for the Australian manufacturing industry as a whole.

After Electrolux took over in 2001, this secure culture was ruptured. Workers encountered non-negotiable, continuous restructuring which created profound insecurity. In addition, the overwhelming majority felt that work was being intensified. As one worker commented: 'We are being pushed a lot more to do our jobs faster. When people work faster there is always confusion and stuff-ups and sometimes even accidents. We are humans, not robots'.[31] Another said: 'We are having to do multiple jobs.'[32] In making a reference to the takeover of a factory from the previous owners Email by Electrolux, another mentioned: 'We have had to work harder over the last six months, ever since our old boss left and the new one took over.'[33] For the first time in the factory's history, management stated repeatedly that the plant's future will not be guaranteed (Lambert, Gillan & Fitzgerald 2005). Feelings of anxiety were exacerbated by the corporation's embrace of the new anti-union industrial relations regime introduced by the Howard-led conservative government.[34] The application of these laws, which created ideal conditions for restructuring, are briefly elaborated as they consolidated the structural domination of capital.

Within 18 months of the takeover of what was previously known as Email, Electrolux shattered the Orange experience by reducing the 1,800 workforce to just 750 through continuous downsizing, outsourcing, and work intensification. Jenny, an assembly line worker, stated: 'I'm over fifty. I have no skills and I'm frightened. I've got kids growing up in the town. We want Orange to go ahead, but we feel insecure now. We are unsure of the future and we're damned frightened'.[35] During this period the company demonstrated its command over space by first outsourcing a number of manufacturing operations to local suppliers. When these suppliers were unable to compete with the 'China price', contracts were shifted to overseas suppliers, increasingly rendering Orange little more than an assembly operation rather than a manufacturing process.

The company developed a carefully planned strategy to counter possible union resistance to these changes. Shop floor unionists were labelled 'trouble makers' who were unable to adapt to the rigours of global competition. Only those who cooperated and adjusted could be sure of (unspecified) job security, while recalcitrant workers advancing an 'adversarial culture' were the first out the door. A worker observed that the culture of challenging management was crushed during this period.

> Now there are a lot of yes-men around, with brown noses and brown tongues, who won't stand up for anything.[36]

Restructuring, insecurity and cultural subordination have led to union decline, with density down from 95 to 70 percent. Furthermore, management used restructuring to reconfigure the age profile of the workforce, replacing older with younger workers who are scared to participate in union events, fearing they 'might be sacked'. Since the takeover, union meetings have generally been poorly attended. This has been exacerbated by the company's anti-union strategy. Shop floor delegates are not given time to speak to members. One commented, 'I took three minutes off

the line to speak to a young worker about joining the union. The fore-
man comes along immediately and says – "You should be working!"
It's difficult to recruit the new workers in conditions like this.' In fact,
like the early days of the Ford Motor Company, workers in general are
prohibited from speaking to each other on the line as such conversation
might adversely affect speed (Beynon 1973).

Most significantly, Electrolux has seized the opportunity of the federal
government's 1996 Workplace Relations Act to further marginalize
unionism. Right of access and secret ballot provisions of the Act were
applied to erode any lingering solidarity culture. Union organizers have
been denied right of entry since May 2003 and the ballot process,
designed to individualize workers' response to workplace issues, has
been relatively successful. Electrolux applied this measure during enterprise bargaining negotiations, going to a ballot when negotiations dead-
locked. Failure to win the first ballot by a narrow margin in June 2003
led to a re-instigation of the process in July, which the company then
won by a substantial margin.[37] The law allows for those not directly
affected by the provisions of the new agreement to vote. This included
managerial, office and engineering staff, leading an organizer form the
AWU to view this 'as a first step in a process to do away with unions in
the 65-year-old factory'.[38] Denied right of entry, union organizers were
marginalized in the bargaining process as they were prevented from
articulating an alternative position prior to the ballot. Provisions under
section 170LK provide employers with further options to force unions
to the periphery of bargaining through enabling non-union collective
agreements. Although not used, the non-union LK threat was ever
present with senior management openly expressing their desire for 'company unions' or 'in-house tribunals'. Throughout this process, management heightened fear and insecurity by emphasizing the uncertainty
inherent in Electrolux's global production strategy, claiming that the
long-term viability of the plant was far from guaranteed, hence workers
needed to demonstrate that they were willing to make concessions.

In June 2006 the union agreed to remove a clause in their collective
agreement that required the firm to appoint casual workers on a permanent contract if they worked for longer than six months at the
factory. The union did, however, win the right for casual workers to be
paid at the shop rate of $22 an hour, rather than the award rate, which is
much lower. Casual workers received no additional benefits and in
November 2006 the union delegates estimated that there were 100
casual workers working in the factory in addition to the existing 350
employees on standard contracts of employment. In fact, on the day we

had this meeting with the delegates, the company brought in an additional 20 casual workers. They could be dismissed with an hour's notice and were sourced by a recruitment agency. However, they were directly employed by the firm. These were mostly younger workers, often students who wanted to make a quick buck.

While workers were bargaining their rights away in Orange in the hope of maintaining the plant's presence, their jobs and their rural lifestyle, Electrolux further advanced its global strategy by investing in its Chinese 'global production platform'. The company's spatial fix is a US$50 million new investment of upgrading its Changsha factory in the inland Hunan province. Fridge production expanded from 650,000 per annum to 1.3 million by 2006, a move which has obvious implications for Australian workers. In May 2004 the corporation cut another 400 jobs, slashing the workforce to a mere 350, thereby further consolidating the factory's status as an assembly rather than manufacturing operation. It appears that its commitment to Orange is now a short-term one, while the Changsha operation is fine-tuned. Once the China plant is fully operational, fridge assembly in Australia is likely to be rationalized out of existence. The unions have been informed that they will have no say over redundancies and they are conscious of the strategy of targeting unionists first.

The case of Electrolux in Orange illustrates the shift from a hegemonic labour regime, where state-supported industrial relations procedures allowed for significant influence of unions in the workplace, to one of growing insecurity in the face of international competition. This has led to the emergence of concession bargaining in a desperate attempt to keep the plant open, a classic example of what Burawoy calls hegemonic despotism. The result has been a weakening of the unions, accelerated by the company's use of measures provided for by labour law reforms. With the new labour law reforms allowing for individual contracts of employment, hegemonic despotism has become full-blown market despotism (Peetz 2006).

Ezakheni: You Can Rely On Defy . . .

Defy's refrigerator factory in Ezakheni was custom-built for the company by the KwaZulu Finance Corporation in their industrial park in 1986. This was at the height of the investment boom in the area, when they 'built a factory a day'. Still, Defy's decision to locate its operation here was considered a 'lucky break' by local state functionaries.[39]

Defy is considered to be a stable investor because it is locally owned and it pays its workers much higher wages than most of the surrounding Taiwanese, and increasingly Chinese, textile and clothing operations in the area. The Ezakheni industrial park is located in the former KwaZulu, an area designated by the apartheid regime as an ethnic 'homeland' exclusively for Zulu people, and the location of the plant in this area was an attempt at a spatial fix in response to urban militancy. Firms took advantage of the apartheid geography to escape union recognition in urban areas. As we shall see, such a fix is never a permanent solution, as space is also contested by workers.

Defy's main manufacturing operation is located in Jacobs, in the south of Durban, South Africa's major port. Indeed, this is one of the oldest factories in South Africa. Whereas workers at the Defy factory in Jacobs were able to organize themselves into trade unions in a less repressive legislative environment after the changes to labour laws in 1979, workers from places like Ezakheni 'still toiled without rights and protection in the "homelands" ... places Wiehahn[40] never touched' (Friedman 1987: 475). As Friedman points out, they were 'the workers the eighties forgot':

> Their plight was usually ignored, for they were conveniently out of sight. But it remained no less important because it revealed how much had not been changed by the reforms ... Most labour laws did not cover these areas. In 1970, the government had exempted them from the Industrial Conciliation Act: it hoped that, by freeing firms from the minimum wage clauses in industrial agreements, it would help 'homeland' industry grow and provide the jobs to stem the flow of Africans to the 'white' cities. (Friedman 1987: 475)

However, this internal spatial fix – formerly known as industrial decentralization – pre-dates formal apartheid. In the 1940s, as part of the Smuts government's import substitution industrialization programme, '[m]oves in this direction' had already begun. Hart (2002: 134) argues that the state was confronted by increased militancy among African workers in urban areas who had begun to move into 'skilled operative jobs reserved for whites' (Hart 2002: 134–135).

The 1970s was a decade when the democratic labour movement emerged and the state was forced to reconsider its approach to industrial relations. 1973 was a turning point, when a spontaneous strike wave started in Durban and Pinetown. However, throughout the 1970s and the 1980s, factories located in Ezakheni and other industrial decentralization zones were particularly hard to organize. These companies had the advantages of exemptions from minimum wage requirements, generous government subsidies, and a repressive regime that violently opposed attempts to organize workers into democratic unions. Defy's location in 1986 of its new refrigerator factory in Ezakheni should be seen in this context. Nevertheless, the emerging metal union, the National Union of Metalworkers of South Africa (NUMSA), an affiliate of South Africa's largest federation, the Congress of South African Trade Unions (COSATU), began organizing the plant. It was a time of heightened militancy among workers, including struggles in the communities over forced land removals. Importantly, NUMSA was a national industrial union with an organized presence throughout South Africa, including Defy's other plants in Jacobs and East London in the Eastern Cape. The plant in East London was also located in a former homeland, the Ciskei, known pejoratively as a Bantustan. With the transition to democracy, and the formal incorporation of these areas into South African labour law, the union branches increasingly linked up, and formulated a company-wide bargaining strategy (Bezuidenhout & Webster 2008).

However, the end of apartheid did not mean that industrial decentralization zones such as Ezakheni withered away. The KwaZulu Finance Corporation reinvented itself as Ithala Development Finance Corporation Limited, which still plays an active role in promoting industrial parks, and makes deals with local government to supply its clients with access to low rent and services. Despite these attempts to maintain incentives for companies, many firms have relocated since the end of apartheid. With numerous closures of plants, the park has become more dependent on Defy for its prestige as a desirable site for manufacturing (Bezuidenhout 2006).

Defy's shift of its Kelvinator operations from Johannesburg to Ezakheni in 1999, as well as its acquisition of two production sites from Silu and Kheni refrigeration, an Asian-owned operation in Ezakheni, has meant an expansion of production capacity. However, this has not translated into an increase in secure employment. Although labour market segmentation is less elaborate than in Changwon, factory workers are segmented into a core group of permanent employees and a second layer of short-term contractors. These workers are popularly referred to as STCs. The firm wanted to introduce labour brokers to source casual workers, but the union successfully resisted these attempts. Defy's STC workers generally work for the latter six months of the year to take advantage of the Christmas market. This adds to the sense of insecurity. A worker described the system as follows:

> What they do, if the contractor has worked, say for six months, they will terminate the contract. And then they would say: 'No, they don't have the job for the contract anymore.' All of a sudden, after two or three weeks, they will say they need those contractors back again. You see, [that is] the strategy they are using.[41]

Another explained the market rationale:

> I think the company doesn't want to pay. That's why they employ contractors ... [I]t's cheaper, because when they don't want them again they just tell them: 'Your closing date is in June. We are taking you away. We don't want you anymore.' It's easy for them – rather than [with] permanent [workers, where] they have to go through some stages.[42]

An interviewee even went so far as to suggest that 'the only difference [between] contract and permanent [workers] I think is being a member of the union – having a card and everything; that's the only difference.' According to her 'everything [else] is just the same'.[43]

Older permanent workers tended to be more critical of management than younger STCs. While these workers were committed to their union, they had a sense of it losing influence in the factory. According to a 31-year-old, permanent male employee, their relationship with management had deteriorated. 'In 2000 we were recognized as a workforce and NUMSA. Now we communicate through the shop stewards.' He added, however, that 'workers' demands [were] often rejected'.[44] Another permanent worker (40 years old) who had worked at Defy for 15 years made a very strong statement: 'It is worse than before. People are treated like animals. Management dictates even in issues such as safety. The union (NUMSA) is helping a great deal to make the job bearable.'[45]

These generational tensions that intersected with the segmented labour market were also apparent. A young, female contract worker explained:

> Especially contract workers [are] younger people at the age of twenties ...
> And permanent people are the old people who've been here for ten years, and everything. And it's like when you start complaining maybe about the wages that you get, you know, maybe you say [with] this money you can't do anything. And they [the older workers would say:] '[When] we started

working here, [we were] earning R75.00 per week. And here you are earning R400.00 per week and [you] still complain!' And we [would say:] 'Like it's not our fault!' I mean, it's not a lot! [A]nd you get people who've been here for ten years and they are still in grade-F, you know? And there are people who've been here less than a year and are on higher grades than them. Then this issue just [creates conflict].[46]

This conflict was aggravated by the fact that the different generations had different levels of formal education. The interviewee pointed out that, of the batch of 30 contract workers who were employed by the company when she got her job there, only two did not have some form of tertiary education qualification. Most had diplomas (mostly teaching diplomas), and two had degrees from universities. She saw this change in hiring practice as having changed over time:

The ones [who were employed] before us had matric. But the older ones, they don't have matric or formal education, which makes it another issue, you know? ... [I]t's like we are here – educated, but still earning the same money as they do ... [T]hen they start making jokes and everything, and then it just gets tense ... [So] I think there's another problem which needs to be looked at ... [T]he older workers see us as a threat, you know; taking their jobs. They started bringing up this issue of them having longer service years than us ... so they [say that they] deserve to get better jobs.[47]

The tension between older and younger generations of workers in this factory is expressed as experience versus education. Older workers emphasize the fact that they have worked in the factory for a long time. Younger workers feel the older ones are not as educated as they are – this is a legacy of the apartheid education system. These cleavages, which coincide with the company's internally segmented labour market, create tensions among workers and divisions in the union. We see in this strategy the further segmentation of the workforce and the community through the manipulation of generational and educational difference.

Workers at this factory, as well as Defy's two other factories, were still campaigning to unmake the legacy of the apartheid workplace regime. An issue that led to a company-wide strike in 2002 was what workers called 'grade anomalies'. The company had moved to a new grading system, and it transpired that workers appointed in similar grades got different salaries. Those who got higher wages often happened to be white. The union saw this as discrimination. It also challenged the firm on its training policies, as well as its perceived lack of commitment to redressing racial inequalities in the workplace through employment equity measures. Workers generally viewed the firm as authoritarian and felt that management stuck to a mindset of the past (Bezuidenhout 2004). A 26-year-old female contract worker mentioned that she believed 'there is still some discrimination, as most supervisors and superintendents are Indians and whites'.[48]

Workers in the area constantly live under the threat of plant closures. Indeed, when interviewed in 2001, retrenchments at the factory in 1997 and 1999 were fresh in their memories.[49] They were concerned also about the status of the Defy factory: 'They are ... saying this company, Defy Ezakheni, hasn't made profit since it was established here – they are still running at a loss.'[50] However, while concerned, some interviewees were sceptical about the truth of such claims:

> They are making a profit. If they were ... not making [a profit], they wouldn't have [bought] the assets from Kelvinator ... We are making our own condensers now. Previously we were buying condensers from outside, but now we are making our own condensers. So to us, we are not convinced [when] they are saying we are not making profit.[51]

In fact, Defy has succeeded in capturing an export market for chest freezers, mainly for the African market, which are manufactured at its plant in Ezakheni (Mohammed & Roberts 2006).

A further source of insecurity at this factory is the fact that there are day and night shifts in some departments. Some employees were unhappy because 'there's no transport at night'. Apparently, when the night shift was introduced, workers were promised transport at night. However, 'when the people agreed to work shift, there was no transport at all.'[52] These concerns should be seen in the context of high crime levels in Ezakheni and the fact that workers are concerned about their personal security (see chapter 7).

In Ezakheni, workers linked the intensification of work to a number of different factors. Some mentioned the cyclical nature of production at this factory. Others felt that new machinery had led to intensification. An older (51-year-old) permanent worker felt that there were too few workers with too little skill. A number of our interviewees said that they did not know much, or did not care, about their relationship with their management. 'I don't really know. I'm not that interested in these issues as long as I am working. I've spent too many years being unemployed, or being paid peanuts.'[53]

The establishment of the Defy factory in Ezakheni in 1986 was an attempt to maintain the colonial despotism that was challenged by workers in Durban. When this area was incorporated into the South African labour regime in the 1990s, the union drew on the reformed labour regime to establish its influence in the workplace. We see the beginnings of a hegemonic regime. However, this is soon challenged by management's introduction of a layer of STC workers, who reintroduce workplace insecurity. While Defy seems committed to keep the factory in Ezakheni, workers feel insecure because of extremely high levels of unemployment and the closure of other factories in the vicinity. Hence, we see the double transition leading to a labour regime that approximates market despotism.

Conclusion

LG, Electrolux and Defy run three different factories in three countries with diverse histories. Yet a common theme emerges when we consider the workplace regime in each of these three factories. Workers generally feel insecure – they are anxious about the possibility that they may lose their jobs, or that the content of their jobs may be altered to the extent that their lives become increasingly insecure. The threat of relocation looms large, work is intensified, and the segmentation of the labour market adds a significant disciplinary element. Hence, at each of the three sites, workers are aware that they cannot take their places of work for granted. This is used by management to extract concessions from them. The segmentation of the labour market into insiders and outsiders is a constant reminder to those who are lucky to have more stable employment that they may be downgraded. In both Ezakheni and Orange, employers are recruiting younger workers, which further segments the labour market and challenges longstanding union traditions. If you are in the core of the labour market, you work hard, because you know you may end up on the periphery. If you work on the periphery, you know that your position is tenuous in the first place. In Korea this has become a major social issue – since more than half the labour market can now be classified as 'irregular' (Yun 2007). In South Africa workers accept the need for short-term contract labour, and the union does not even bother to organize them. In Australia Electrolux now employs a segment of younger casual workers in addition to workers on standard contracts of employment. The firm has also outsourced certain parts of the production process to its factories in China. This has dramatically reduced the size of the workforce. All these elements contribute to an overwhelming sense of insecurity and a decline in workers' bargaining power.

In the case of Changwon, LG's suppliers are increasingly moving to China. Other Korean white goods manufacturers have already decided to move production elsewhere. Hence, while LG is still committed to remaining in Changwon, workers have to work much harder in order to keep their factory there. They are constantly made aware that their wages and the strength of their currency agitate against this commitment from management. Irregular workers, who are not union members, and the peculiar industrial structure of LG's operations, undermine the workplace bargaining power of workers. Since the financial crisis of 1997, Korean workers are more concerned about issues of unemployment. Since welfare is linked to the company, and not the state, they do not have significant marketplace bargaining power. The presence of an entrenched company union prevents LG's core workers from developing a democratic union and hence strong associational bargaining power. As was illustrated, workers in LG's supply firms are too scared of associating openly with the federation of their choice, the KCTU. This further reflects their lack of associational power.

Since Electrolux's takeover of the factory in Orange, jobs have been decimated. A stable manufacturing operation was reduced to a mere assembly plant. The company shows no commitment to maintain even the skeleton the factory has become. Workers who remain have become cynical and scared. Those who speak up are targeted for the next round of lay-offs. Such dramatic lay-offs and the very real possibility that the factory may close down altogether, have weakened workers' workplace bargaining power. In this context, even though Australia has a lower unemployment rate than say South Africa, workers can draw on very little marketplace bargaining power. The targeting of union activists, as well as new labour laws, has also undermined associational bargaining power.

Historically, Ezakheni is a low-wage area. The plant is located here because of its cost competitiveness and state subsidies. But these subsidies have now come to an end and the fact that firms in the area constantly close down to move elsewhere reminds workers that the same can happen to them. Although workplace bargaining power has been weakened somewhat by the existence of a layer of STC workers, a more significant challenge to workers' power resides in high levels of unemployment. This limits the marketplace bargaining power of workers in Ezakheni. However, their associational bargaining power is still significant, since the union mobilizes against the remnants of the apartheid workplace regime.

In the age of insecurity, the factories in Changwon, Orange and Ezakheni do not only manufacture refrigerators, they actively manufacture insecurity as the key source of discipline over workers. Some elements of hegemony are employed – especially for those in more stable employment – but the defining feature of the working lives of these workers is the fact that they are insecure. There is a convergence in all three workplaces towards market despotism.

We turn now to an examination of how market forces have been extended into essential services in all three countries.

4

Citizenship Matters

Household refrigeration is basically about food preservation. If you can buy meat, vegetables or milk in bulk, you benefit from economies of scale and save travel costs. While doing research in Ezakheni, we were asked by one of our informants to assist her with a form she had to fill in. It was getting dark as we were about to complete the paperwork. We had to strain our eyes to read, and we asked her if she could switch on the lights. She first had to find a pre-paid card, and then punch numbers into the electricity unit. In Ezakheni, you use electricity on a 'pay as you go' basis. You leave the lights off for as long as you possibly can, so that your refrigerator can run for the whole month. That is if you have enough money to do this. Very few households have luxuries such as washing machines. Indeed, only 39.3 per cent of Ezakheni households use electricity for heating purposes and 43.1 per cent for cooking and 67.8 per cent for lighting (Emnambithi Local Government 2002: 6).

This chapter is about the politics of electricity and water provision. In the previous two chapters we have seen how insecurity is created by corporate and workplace restructuring. We now show how privatization creates household insecurity.

Contemporary politics in South Korea, Australia and South Africa are defined by market ideology, a key feature of which is the embrace of privatization. In differing degrees, the three nations have privatized electricity and water provision, needs that contribute to defining the quality of everyday existence. However, restructuring essential services is not simply a question of technical service delivery; nor can outcomes be evaluated narrowly on price. Acquisitions of essential services reflect a market-driven politics characterized by the penetration of corporate power into the activities of the state, which erodes democracy, citizenship and the public interest. In short, the political orientation inherent in this ideology needs scrutiny, since the evaluation of this strategy determines the character of a civil society response. Our comparative perspective reveals a range of responses from passive acquiescence (Australia) to varying degrees of resistance (South Africa and South Korea).

We analyse two facets of this policy shift: the nature of the privatization process in each country and its impact on society – where we focus in particular on how privatization shapes the character of democracy and affects households. In Australia we take a detour from our focus on New South Wales, the home state of Orange, to the neighbouring state of Victoria. Of our three cases, privatization has gone furthest in the state of Victoria where the sale of electricity assets was initially secured by a Californian TNC, Edison Mission Electricity (EME). Here, privatization has eroded democracy without bringing any economic benefit to citizens. In South Africa, privatization has exacerbated household insecurity and has led to resistance, while in Korea the privatization of electricity was successfully blocked.

Electricity Privatization in Victoria, Australia

The privatization of electricity in Victoria is shrouded in secrecy, ensuring that citizens have no knowledge of the process, since the process privatized the state from within. The strategy was introduced by the Australian Labour Party (ALP), revealing the remarkable political transmutation that privatization heralds. Historically, social democratic parties mediated the conflicting interests of corporations and civil society, striving to strike a class compromise that accommodated corporate interests to a degree, while also representing a wider public interest. The ALP was a classic social democratic party that struggled with these contradictions. Before privatization, Australian public services such as electricity were provided by both public and private initiatives. However, through the early decades of the twentieth century, under the influence of the ALP, the state extended the provision of public goods and services as part of the construction of a dynamic social democratic state, asserting the needs of society over market logic.

Between the 1890s and 1920s the Australian left were influenced by the Russian revolution. They attempted to extend these ideas to Australia, advocating a nationalization strategy that expanded and protected public services. The nationalization agenda initiated by the Melbourne Trade Hall Council was adopted by the Victorian Labour Conference in 1920 (Turner 1979). Based on a guild socialist approach, the demands included

> the nationalization of banking and the principal industries, control of nationalized industries by boards on which both workers directly involved and the community at large would be represented, and the establishment by these boards of a Supreme Economic Council which would plan and co-ordinate the whole of the national economy. (Turner 1979: 220)

These proposals were debated within the labour movement and more moderate arguments prevailed. Nevertheless, a proactive social democratic vision still held sway in a nation that had become 'a pioneer in progressive social laws which meant, in effect, the advancement of state power' (Kelly 1992: 10). Kelly observes that this reflects a political tradition in which 'Australian democracy has come to look upon the State as a vast public utility whose duty is to provide the greatest possible happiness for the greatest number ... to the Australian, the State means collective power at the service of individualistic rights' (Kelly 1992: 10). Kelly recounts a 1920s observation of a Victorian state minister who in speaking of Victoria's public services noted that, 'in proportion to the size and the economic standing of the community, Victoria's public services constitute possibly the largest and most comprehensive use of state power outside of Russia'. This included intervention in banking, railways, roads, water supply and electricity (Kelly 1992: 11).

Kelly's account of the politics of the 1980s in Australia is a market critique of this 'state socialism'. For Kelly, the 1980s is a period when an 'obsolescent old order' gives way to the new political ideas of the market that 'remade' the Australian political tradition and offered a new definition of nationhood. The ALP was transformed in the early 1980s, ushering in a new era where both major parties mirrored market politics. Both believed in 'the power of markets' and the internationalization of the Australian economy. A less interventionist state would overcome 'market distortions' thereby creating competition, which would 'foster a more dynamic citizenry' and a culture of individual enterprise (Kelly 1992: 11).

This seismic political shift within the ALP was led by Prime Minister Bob Hawke, ex-President of the Australian Council of Trade Unions (ACTU), and Paul Keating, his Treasurer. On coming to power in 1983, they aspired to create a new Labour Party, one with strong links to the business and financial community that would espouse new values that challenged Labour's orthodoxy. These were market values – a belief in the dynamism of free trade, competition, privatization and a radically reduced role for the state. The party leadership was connected to 'the big end of town – the centres of corporate and financial power' and they accepted the prescriptions of the IMF and the OECD (Kelly 1992: 23). The leadership was willing to 'abandon its own internal resources – the dwindling advocates of socialist theory – as the source of its economic policy direction. It wanted to follow the experts and the experts were a new generation of Canberra-based economists commanding the senior posts in the major policy departments who believed in the efficiency of markets and deregulation' (Kelly 1992: 24; Pusey 1991). This is viewed as part of a much wider sea change in which social democratic parties internationally admitted 'that the power of the state to deliver prosperity and justice to its citizens was failing' and that the only solution was 'the embrace of market-based solutions' (Kelly 1992: 31; Sassoon 1996: 730–755).

This embrace of market ideology led to an extensive wave of ALP-driven privatizations initiated in 1985. Promoted by the IMF, the World Bank and the OECD, this restructuring synchronized with privatizations across the globe where the targets were electricity, gas, telecommunications, transport, water and sewerage (table 4.1). Electricity was the second largest revenue earner, after telecommunications (Izaguirre 2002). Nearly 40 per cent of privatization revenue in Australia came from the sale of state electricity and gas (Brune, Garrett & Kogut 2004).

The Labour government privatized so vigorously after 1985 that by 2002 Australia had transformed the state from being the most

Table 4.1 Total Revenue (1990–2000) from privatization

Sector	US$ billion
Telecommunications	331.4
Electricity	213.3
Transport	135.3
Water and sewerage	39.7
Gas (transmission and distribution)	34.5
Total	754.2

Source: Izaguirre (2002)

interventionist to being the most privatized in the OECD after France. According to the OECD, privatization proceeds in Australia over the 1990–2001 period totalled US$ 69 billion – compared to France (US$ 74 billion), UK (US$ 43 billion), Japan (US$ 38 billion), Germany (US$ 25 billion), Korea (US$ 15 billion), Canada (US$ 11 billion) and US (US$ 7 billion) (OECD 2002). In Australia, the average revenues from privatization from 1985 to 1999 amounted to 25 per cent of GDP (in 1985 terms), compared with 28 per cent in Malaysia, 27 per cent in New Zealand, 15 per cent in the UK, and around 5 per cent in Canada (Brune et al. 2004). The extensive sell-off of state assets to global corporations simply extended the high degree of foreign ownership of all key economic sectors that has characterized the economy since the 1970s, a phenomenon labelled 'The Selling of Australia'. Crough and Weelwright observe, 'today, for all practical purposes, Australia has been sold. There are, it is true, a few juicy portions still left ... but most of the really profitable areas have already gone. Australia now has the highest level of foreign ownership and control of all the advanced countries of the world, except Canada' (Crough & Weelwright 1982: 1).

The Australian political system is based on a federal state that functions in collaboration with state governments. In the initial phase of privatization, the federal government took the lead, triggering similar moves by state governments (Fairbrother & Testi 2002: 8). When the Liberal-National Coalition came to power in 1996, it extended and radicalized privatization through 'the opening of state services to tender and franchise' (Fairbrother & Testi 2002: 10). This new phase of privatization assumed the form of Private Public Partnerships (PPPs). This 'partnership' took the remarkable step of transferring all risk to the state and all profit to the corporation acquiring the asset (Sheil 2002; Leys 2001).

The architects of these changes claim economic gains justify the strategy. While our data on the electricity privatization of Victoria contradict this claim, our critique extends to the manner in which corporations penetrated state apparatuses, facilitating a direct influence in internal decision making, creating the 'captive state' (Monbiot 2000). The creation of a National Competition Policy (NCP) became the instrument for furthering this profit and political influence opportunity for private corporations.

NCP provided the legislative architecture to advance privatization (Ranald 1995). The NCP coincided with the mutation of the General Agreement on Tariffs and Trade (GATT) into the WTO in 1995, and the emergence of the GATS (General Agreement on Trade in Services), opening public services worldwide to takeover by global corporations.

The NCP shaped the privatization process through defining core and non-core activities of the state and then advocating the privatization of non-core activities through a process of contracting out. In the 1990s the Liberal Kennett government in Victoria required local governments to contract out services such as garbage collection, recycling, cleaning, the creation and maintenance of parks and gardens, the running of meals-on-wheels, and library services. In these processes there is a critical weakening in the public accountability of the state to the democratic process (parliament) and to civil society (trade unions, interest groups and social movements).

In addition, government audits have been privatized. Nothing highlights the captive state more than this erosion of sovereignty in the takeover of public sector accounting and auditing by global and domestic corporations. Public accountability has been incalculably compromised (Karan 2003). The implication is that privatizations are secretly struck deals where the public right to know is marginalized under 'commercial in confidence' provisions, which limit public disclosure of transaction costs and preclude effective monitoring of privatization. This pre-emption of democratic accountability is now standard practice. For example, in 1997, the Howard government outsourced government information technology services. When the Auditor General decided to scrutinize this process in September 2000 the government countered by appointing advisors from large accounting companies as 'independent consultants' who would carry out this role (ABC 2000).

In the context of this changed state role we turn to the privatization of electricity in Victoria.

Electricity Privatization and Public Accountability

The privatization of electricity in Victoria is a clear example of the erosion of democracy and the disempowerment of civil society. The NCP mechanisms cut parliament out of the process and excluded the voice of civil society. The story of this privatization highlights how NCP renders global corporations unaccountable, guarded and protected as they are by secret contracts that not even the Auditor General can scrutinize. Members of parliament are not provided with answers to their questions. Accountability withers before a state which defends corporations against society.

Since the industry was established in the 1920s, electricity was state owned, with the exception of only a few private ventures (Booth 2003;

Johnson & Rix 1993). The privatization of electricity began in earnest in 1991 when electricity service provision was divided into discrete units (Strategic Business Units): generation (power plants), transmission (the high voltage grid) and distribution (poles and wires) in the name of creating competition. NSW and Victoria spearheaded the implementation of these NCP electricity reforms, which included the establishment of a national electricity market (NEM).

The restructuring of the State Electricity Commission of Victoria (SECV) was initiated in the late 1980s with the planned privatization of the four Victorian electricity generating plants. These assets were first corporatized and then sold to US and UK global corporations who formed a consortium (Fairbrother & Testi 2002: 13). Table 4.2 shows the structure and the corporate entities that they created.

The Kennett coalition government in Victoria privatized electricity between 1992 and 1998. The modelling was developed by two right-wing think-tanks, the Institute of Public Affairs and the Tasman Institute (Hayward 1999: 141, 152). The negotiations between the government and specialist private consultants who were mandated to realize privatization were conducted under conditions of intense secrecy (Hodge 2003; Booth 2003). The Electricity Supply Industry (ESI) Reform Unit, which was set up within Treasury, pursued its mandate to transform the SECV ruthlessly, neutralizing opposing politicians, senior public servants and senior managers of companies in the industry (Booth 2003: 50).

Table 4.2 Structure of the consortium that privatized the State Electricity Commission of Victoria

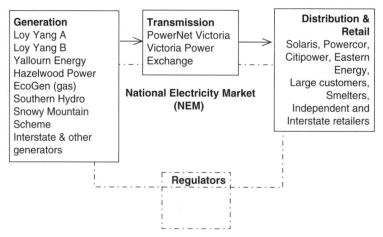

Edison mission energy

Only one Australian company succeeded in purchasing energy assets. Energy Brix bought out the only coal-based utility operating in the LaTrobe valley. All the other state-owned assets were acquired by global corporations, which would remove inefficiencies in the national electricity grid, according to the think-tanks (Fairbrother, Paddon & Teicher 2002: 107).

The secrecy surrounding these deals made it difficult to comprehend the complex web of ownership in the takeover. In 1994 Edison Mission Energy, a subsidiary of California-based Edison International, bought the coal-fired Loy Yang B plant. EME contributed 25 per cent of Edison International's revenues in 2004. Clearly, buying into electricity assets generates high revenues, for in 2003 these were US$12.1 billion, which produced a net income of $821 million. In 2003 the company employed a staff of 15,407 (Hoover's 2004).

The record of EME is pertinent. In September 1996, EME successfully lobbied for the deregulation of electricity markets. By June 2000 the wholesale price of electricity continued to increase, contrary to the expectations of deregulation advocates. In January 2001, SCE, a company linked to EME, defaulted on $569 million worth of payments to power companies and bondholders. The same month the first rolling blackouts since the Second World War hit California, followed by a second major blackout two months later. Following the bankruptcy announcement of PG&E in April 2001, the California governor arranged to bail out SCE in return for the transfer of transmission lines back to the state (FTCR 2002a). In absorbing risks and the costs of privatization, the California government bailed out the very companies that manipulated the electricity regulations to maximize their profits (FTCR 2002b). The rest of the world was not immune from these issues, as EME has interests in Europe and the Asia-Pacific region. In early 2004, EME owned around 80 power plants outside the US and its Asia-Pacific operations included the Philippines, Indonesia, Thailand, New Zealand and Australia. When EME purchased Australian electricity assets in 1994, it was merely one facet of this global empire and Victoria would not be immune from these Californian crises.

When EME faced bankruptcy in 2004 it offloaded its foreign assets, selling 14 power generation companies across Europe, Asia, Australia, New Zealand and Puerto Rico for US$7 billion. In December 2004, EME's Australian assets were bought by UK based International

Table 4.3 EME network in 2003

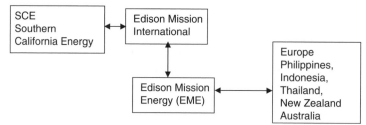

Power (established from the UK's privatized Central Electricity Generating Board in 1990) and Mitsui & Co., one of Japan's two leading trading companies (Clifford 2004). There is no public assessment of these deals and their cost to Australian taxpayers. Parliament and civil society remain on the margin. Nor has privatization delivered any economic gain for citizens. After nearly a decade of privatization, residential customers are paying 3 per cent more on average for electricity, and for some customers the increase is as high as 24 per cent (Sharam 2003). According to the Essential Services Commission, by 2002 the average electricity bill decreased compared to 1994/1995 levels (ESC 2003). However, this involved lower average bills for farmers and business customers, while household bills have increased since 2001. In 2002, electricity disconnections increased and the complaints regarding affordability increased 78 per cent from 2001 (ESC 2003). Apart from the increasing average retail price since 1998, privatization has also increased costs of transmission and distribution (Hodge 2003). The impact on the environment has also been negative, with increased generation from brown-coal plants and a pause in the use of gas for power generation exacerbating green house gas emissions (Hodge 2003).

The ACTU remains trapped on the issue of privatization, reflecting its subordination to the ALP, which was the architect of privatization in Australia. At present the ACTU's position is to insist on a 'robust public interest test':

> The ACTU will work with affiliates to achieve government support for a public interest test that should form the basis of quality public services ... Any public interest test must incorporate issues of affordability, accountability, transparency and quality of services. (ACTU 2003)

Such a position ignores the politics of privatization, which erodes democracy and the public interest and marginalizes civil society. As a consequence of the ACTU's weak position on privatization, civil society is muted. And the absence of an alternative discourse has consequences for Australian politics more generally.

Essential Service Privatization in South Africa

The South African case is much broader than just electricity, with a strong emphasis on water and sanitation. In stark contrast to life under social democracy in Australia , South Africa's black population experienced under apartheid an absence of political rights, as well as other basic human rights. This included land dispossession, family disintegration through migration policies that separated families, the absence of organizing rights at work, and below-subsistence wages. Work, wages, housing, access to water and electricity and lifestyle opportunities were racially determined. The pass law system, and influx control measures, only allowed a limited number of black South Africans to move to urban areas. In some 'townships', which operated as labour reserves for urban industry and households, basic subsidized services were provided. These services existed to maintain a productive urban working class; citizenship was never on the agenda.

When the African National Congress (ANC) came to power in 1994 it initially embraced a social democratic discourse with the adoption of the Reconstruction and Development Program (RDP), which was part of the ANC's election manifesto. The state 'must play a leading and enabling role in guiding the economy and the market towards reconstruction and development. There needs to be balance between government intervention, the private sector and the participation of civil society. There must be a significant role for public sector investment to complement private sector and community participation in stimulating reconstruction and development' (ANC 1994: 80).

The RDP policy on privatization was ambiguous. However, in 1995 the policy question was clarified when the ANC government signed a National Framework Agreement (NFA) with organized labour, which endorsed the policy of privatization under certain circumstances. The agreement ushered in the ANC's restructuring programme with the sale of segments of the South African Broadcasting Corporation (SABC), the partial sale of South African Airways and the state-owned telecommunications company Telkom. Within six months of signing the NFA this policy direction was further consolidated when the government adopted GEAR (Growth, Employment and Redistribution) in 1996. In this the government argued that,

> The nature of restructuring, as outlined in the framework agreement, may involve the total sale of the asset, a partial sale to strategic equity partners or the sale of the asset with government retaining a strategic interest. (National Framework Agreement 1996: 19)

At the core of these proposals was the 'restructuring' of essential services such as water, electricity, transport, housing and healthcare. ANC discourse sought to legitimate corporate takeover or partnerships by making a number of assurances. Firstly, the restructuring would be guided by a strong developmental state committed to 'the eradication of crushing poverty'. Privatization was needed to improve the quality of life for the majority of the population. Secondly, society would be protected because impact analyses of restructuring, which consider the costs and the benefits to society, would be undertaken before assets were sold. These included direct costs such as the impact of pricing and employment and indirect costs such as increased unemployment. Thirdly, privatization would create globally competitive State Owned Enterprises (SOEs) that would contribute to South Africa's integration into the global economy and the African Renaissance. Fourthly, the sell-off would advance black economic empowerment through 'broadened ownership, training, procurement and self-management opportunities for black people, women and the disabled through involvement with SOE management and ownership opportunities'.

Table 4.4 Contradictions of essential service privatization in South Africa

Shareholder-driven, profit (tariff rises), cost-cutting focus of global and local corporations	The human needs of civil society
The consumer/enterprise culture	Social citizenship/commitment to social justice
User pays	South African Constitution and the Bill of Rights
Aggressive cost recovery	Essential services are a human right
	Need for cheap, affordable services
	The poor have a right to access water and electricity.

The actual privatization of electricity and water captures the real tensions between market and society as the ANC grapples with the contradictions within its discourse and the logic of privatization.

These contradictions are now explored in relation to the privatization of electricity and water.

Electricity privatization

The South African government formed the Electricity Supply Commission (ESKOM) in 1923 to generate, transmit and distribute electricity in South Africa. This SOE is now a significant energy organization by world standards, ranked eleventh in the world in terms of size, and seventh by sales volume. The company employs 29,845 people (ESKOM Annual Report 2005: 189). The 1998 ESKOM Amendment Act corporatized the utility and created separate entities for generation, transmission and distribution, similar to pre-privatization restructuring in Australia. The Act allows the government-owned utility to part-privatize through the sale of shares to private corporations. The Congress of South African Trade Unions (COSATU) opposed this move, arguing that this diluted the state's capacity to direct future investment. Furthermore, a licence was issued for a second national operator in May 2002. According to a ministerial announcement in December 2004 the following corporations invested in the new enterprise: Eskom Enterprises (30 per cent); Transtel (19 per cent); Nexus Connection (12.5 per cent); CommuniTel (12.5 per cent); Two-Consortium (12.5 per cent); Tata Africa Holdings (representing Videsh Sanchar Nigam Limited, an international telecommunications operator in India) (26 per cent) (ESKOM Annual Report 2005: 77).

The restructuring also sought to grapple with the serious problem produced by widespread poverty. A significant number of municipalities, which are required to collect electricity payments from the communities and then transfer the funds to ESKOM, are deep in debt and fail to meet their social service provision commitments. One of the newly created entities, the fully government-owned ESKOM Distribution Industry, is mandated to work through Regional Electricity Distributors (REDs), which will take over the role of the municipalities with regard to electricity distribution and payment.

This strategy is more sophisticated than the state government in Victoria's complete withdrawal from this essential service. It could be argued that the ANC government's continuing strategic presence in this vital service indicates that market forces reflected in price and profit are subject to government regulation in the interests of serving the needs of poor black communities. However, the government appears to be caught on the horns of a dilemma. Its discourse on the restructuring of electricity is built on contradictory value choices. On the one hand, the 1998 government White Paper on Energy states that energy resources must be developed to provide for the needs of the nation – 'Energy should therefore be available to all citizens at an affordable cost'. The paper states that government should 'promote access to affordable energy services for disadvantaged households, small businesses, small farms and community services'. Restructuring needs to redress economic and social imbalances, hence tariffs for low-income households should be capped at affordable levels determined by the National Electricity Regulator. ESKOM claims that it is committed to universal access for electricity.

These social commitments are juxtaposed with the language of the market as though the one simply complements the other. 'Market-based pricing' is essential; an 'investor-friendly environment' should be created; electricity should develop as 'a competitive market'. Universal access must be 'balanced' with 'financial sustainability'. In the cold reality of policy implementation, the market notion of 'user pays' has become the key feature of the new arrangements. Card-facilitated prepaid meters are the mechanism to implement the policy where the 'users' have to pay in advance. ESKOM installed 300,000 new prepaid meters between 1994 and 1999. There are a large number of old prepaid meters already in the field and several replacement projects are already ongoing.

ESKOM now runs on the principle of full cost recovery. Cross-subsidization is supposed to operate to assist certain categories of users, but the effectiveness of this has been questioned. An average price increase of 2.5 per cent was approved by the National Electricity Regulator (NER) in 2003 for a 12-month period to 31 December 2004. In November 2004 the NER approved a 4.1 per cent average price increase for the 15-month period from 1 January 2005 to 31 March 2006 (2005/06 increase) (ESKOM 2005: 70–71).

Water privatization

Nothing captures the contradictions of privatization more poignantly than the issue of water and sanitation in South Africa. The absence of clean drinking water and proper sanitation in segregated black communities was a hallmark of the widespread poverty produced by apartheid. The scale of the problem is enormous. In 2001, of the 44.8 million people in South Africa, 5 million (11 per cent) had no access to a safe water supply and a further 6.5 million (15 per cent) did not have basic service levels; 18.1 million (41 per cent) did not have adequate sanitation services (2001 Census). The ANC government recognizes that it will be judged on whether or not significant improvements in this area are realized, and in many instances advances have been made to extend services to previously excluded communities. But this runs up against the logic of cost recovery. Hence the emphasis in the discourse is commitment to 'the creation of a basic water and sanitation service to all people living in South Africa'. The Department of Water Affairs' (DWAF) slogan is 'Water is Life, Sanitation is Dignity'. A key focus of South Africa's water services policy is to ensure that the poor have access to adequate, affordable and sustainable levels of basic water supply and sanitation services. However, the government's later embrace of the market model led to a significant shift in the discourse, which materialized in a transformed state role. The DWAF redefined its role from direct provider 'to being a sector leader, supporter and regulator' (Strategic Framework, DWAF 2003: 2).

The new discourse cemented this role. The water services authorities (local government) will be constitutionally responsible for the water provision, but this may be *contracted out* to external water services providers (private corporations); there should be *flexibility* in the type of service provider; these enterprises *sell* water to local authorities (water becomes a commodity like any other commodity, where price is

Table 4.5 Water restructuring contradictions

Right of access to basic water and sanitation	User pays (prepaid meters)
	Credit control policy essential, with 'fair warnings given'
Ownership to reside in public domain	Operations contracted (long-term concessions) to private corporations who are primarily responsible to shareholders before all other considerations
State will increase investment in services over time	
Tariffs must take into account the affordability of water services	Tariffs must reflect all the costs associated with the service
Tariffs must be based on equity and fairness	Water services must be run on a 'fully-recoverable basis'
	Tariffs must safeguard the financial viability of the water service provider

determined by market forces); these arrangements are established via *commercial contracts.*

The state has attempted to resolve these contradictions through various subsidies to local governments to try to ensure affordability. It is committed to subsidizing ongoing operating and maintenance costs so as to ensure 25 litres free water per day per person. Where sustainable, the government encourages water services authorities to increase free water from 25 to 50 litres, which means that you can flush your toilet twice.

Long-term contracts with private corporations are central to this restructuring. Table 4.6 lists the private entities involved in water supply.

Responses to essential service privatization

The contradictions between the needs of society against market logic became acute when the ANC government embraced market strategies through the ESKOM restructuring and through contracting the water provision to private corporations. This led to local resistance, as illustrated in the Eastern Cape towns of Fort Beaufort, Queenstown and Stutterheim, the first municipalities to privatize water services when they signed 25-year contracts with Water Services of South Africa (WSSA).

To secure these contracts WSSA, and the global corporations that they are linked to, promised considerable benefits to the ANC government, the municipalities and its citizens. Private corporations, they said, deliver more efficiently and thereby unburden the state from non-core activities. Local authorities were under pressure from demanding consumers, they faced a payments crisis and workers in these utilities were militant. Furthermore, essential service provision increased the workload of municipal officials. The companies' directors asserted that they would free government from this 'arduous and time-consuming work so that they would be able to develop the region and the community' (Strategic Framework, DWAF 2003: 43). Above all, they promised 'affordable and acceptable world-class services at lower than usual municipal costs', promising an 8 per cent savings in Fort Beaufort and 19 per cent in Queenstown.

Once the new corporate system was functional, reality contradicted that promise. Prices increased markedly. This is not surprising since the ANC's policy on the privatization of essential services is one of full cost recovery. Hence the price included operating, infrastructure and maintenance costs as well as a reasonable profit margin for the corporation. Consistent with market ideology, all subsidies must be

Table 4.6 Private corporations involved in water privatization

Local corporations	Linked global corporations
Biwater: Formed in 1968. SA owned company. 30-year contract to provide water services to Nelspruit, a small city in Mpumulanga Province. First ANC foray into the privatization of water. Turnover in 2002/3 was R179 million. Core activity: water and waste management; water infrastructure investment and operation	**Biwater:** Operates in many African countries
Johannesburg Water: formed by the Johannesburg municipal council in 2001.	
Siza Water: majority owned by Saur; five SA partners. Running deficits	**Saur International:** own 58% of Siza
Water Sanitation Services South Africa (WSSA): Subsidiary of Lyonnaise Water, Southern Africa, linked to Lyonnaise de Eau. Works with local government in 6 provinces where 80% of the population fall below the low-income category	**Lyonnaise de Eau:** French MNC. WSSA 'delegated management contracts'. Involved in Johannesburg and a range of rural towns, serving 5.2 million (13% of SA population)
	Suez: leading global energy, water and waste service provider. Employs 190,000 in 130 countries. Annual revenue, US$4 billion in 2001. Contract with WSSA to provide fresh water and waste services to Queenstown and Stutterheim (rural towns)
Umgeni Water: Services KwaZulu-Natal, also to Ezakheni. Tendering for contracts in SA and internationally	

removed, revealing the true cost of the service provided. In Fort Beaufort water connection fees increased by 100 per cent from R310 to R648 and the town's water tariffs were higher than those of non-privatized country towns. In Reddersburg, a small country town similar to Fort Beaufort, residents paid R62 for 30kl compared to the R90 paid in Fort Beaufort.

Table 4.7   Tariffs before and after privatization, Fort Beaufort (in Rand)

	1994/95	1995/96	1996/97	1997/98	1998/99
Basic water (10kl)	6,10	6,10	22,30	27,00	27,80
Sewerage	–	10,13	30,39	35,00	–
Bucket	–	10,13	19,00	22,00	22,60
Electricity	–	4,50	Prepaid	Prepaid	Prepaid
Refuse	4,50	6,50	20,00	23,00	23,70
All services – township flat rates	10,60	28,00	60,00	72,00	74,00

Source: Fort Beaufort TLC Minutes for June and July of each year

Given the low incomes of households, where many members are unemployed or on low-pay work, not surprisingly these changes in the cost structure of essential services generated a payment crisis. This was not a boycott movement replicating the halcyon days of the anti-apartheid campaigns – people simply could not afford to pay. By February 2000 the citizens of Fort Beaufort owed the municipality R13 million. By June the debt had ballooned to R15 million. The average monthly income was a meagre R600 in households that averaged seven to eight people per home, with half of those being unemployed. In such a situation increased essential service payments, which had increased to 30 per cent of the household income, were now beyond the reach of the average household. Most households are in debt of around R5,000 to the Council – an amount far beyond any imaginable recovery.

Increased indebtedness has led to increased coercion and punitive measures against households through simply cutting off access to water and electricity. Cost recovery is enforced in an aggressive manner throughout the country and citizens are issued with threatening lawyers' letters, legal action, evictions, intimidation and overt violence. For example, the Cape Town City Council issued a leaflet entitled, *Pay and stay – it's the right thing to do*, which stated, 'Action will be taken against those who do not pay – the Council will not hesitate to cut off services and take legal action where necessary. Residents who do not pay will be without electricity or water and will have to pay the additional costs of reconnection fees, lawyers' fees and legal costs. They will ultimately have their houses sold (if they are ratepayers) or be evicted (if they are tenants in a Council house)' (2001). Furniture and even the physical structure of the households' dwelling are removed by private security companies, labelled by citizens as the 'red ants'.

Resistance is met by tear gas, rubber bullets and real bullets, as happened in Tafelsig in Cape Town in 2001 (*Cape Times*, 10 September 2001: 31). Many of these family members had become accustomed to this violent response under the apartheid regime. Many actively supported the ANC in the anti-apartheid struggle. Being confronted by tear gas and bullets imposed by the agents of their democratically elected government is disorientating and disillusioning.

Many households have tried to dodge this destructive scenario by turning to a new class of money lenders. A Church survey conducted in the Limpopo province found that people spent 45 per cent of loans on essential service payments, stating that the increasing costs of these payments were a factor contributing to the need for loans. Loans ranged between R1,000 to over R2,000 and the 30 per cent interest rate deepened indebtedness. The survey highlights this debt trap: 'If someone cannot pay for basic living expenses, they won't be able to repay the money lender. This leads to a debt trap where the borrower is unable to pay the first money lender and goes to another lender or uses the same one for another loan. This creates a cycle of debt and further impoverishment, resulting in the borrower's enslavement to money lending.' A national study found that over 70 per cent of those approaching money lenders are repeat borrowers. Many lose control over their finances in this process when they are forced to surrender their ATM cards and secret pin number to the lender, who then uses the card to withdraw repayments. Money lending is one of the fastest growing trends in the financial services sector and there now exists 30,500 lenders, or one for every 1,500 people in South Africa.

A possible outcome of this process of increasing indebtedness is for persons to retreat into their world of 'private troubles' – shamed, embarrassed and wracked with a low sense of self-esteem where borrowers feel a sense of guilt and self-blame that they can't provide for their children. They are particularly pained by the fact that repaying the debt collector often means sacrificing their children's education. The survey found that this sense of the burden of the debt fell largely on women. Seventy-eight per cent of those surveyed said that the lender was the first account honoured. Many feared violent coercion if they failed to meet this commitment. Increasingly, households are discovering that these 'personal troubles' are indeed 'social issues' and are making this connection by resisting privatization.

Between 2001 and 2006 nationwide resistance to essential service privatization began to emerge. Initially, COSATU and in particular one of its affiliates, the South African Municipal Workers' Union

(SAMWU), took action against privatization in 2001 and 2002. On 30 August 2001, COSATU organized a two-day strike that involved thousands of workers. This was a prelude to more substantial action at the beginning of October 2002 when an estimated 180,000 workers participated in country wide marches. In addition, SAMWU is at the forefront of organizing various local protest actions. Apart from these actions of the organized labour movement the crisis produced by privatizing water and electricity has led to a flowering of civil society activist groups in the townships that have organized resident resistance. These include the Landless People's Movement, the Concerned Citizens Forum, the Anti-Privatization Forum, the Orange Farm Crisis Committee, the Abahlali BaseMjondolo (Shack Dwellers Movement), the Soweto Electricity Crisis Committee and a wide range of local residents' associations.

A survey of the actions of these groups over this five-year period reveals the emergence of an active society determined to challenge the market solution advanced by the government. There is a distinct pattern to these actions, which highlights the clash between society and disembedded markets. These groups articulate common demands in all their actions: free access to clean water and affordable electricity (an end to prepaid meters); an end to all evictions and a halt to electricity and water cut-offs and the intimidation of the poor. Action takes the form of protest marches to Council offices or those of the Mayor. Police intervene with tear gas and rubber bullets (and on some occasions, live ammunition), resulting in running street battles, stone throwing, burning tyres and road blockades. Arrests and court actions result in further protests.

Essential service privatization has alienated citizens from local councils and has sometimes led to direct action against the ANC. In the Durban townships of Sydenham, Springfield and Asherville, a slogan, 'No land, No house, No vote' has emerged. In many local elections, participation has been low. For example, in the Khutsong election of March 2006, a mere 232 out of 29,540 voted. Only 200 people attended a meeting organized by the ANC in which the organization's National Chairperson, Mosiuoa Lekota, spoke. In this case, the dispute was about moving Khutsong from the Gauteng province to the North West province. Residents felt that their service levels would be compromised by the move. In research undertaken in Phiri, a suburb of Soweto, residents expressed indignation at the practices that they were forced to adopt to save water. 'People delay flushing the toilet until several people have used it. Pre-paid water is forcing us to live like pigs', one respondent remarked. 'In the olden days people used to care for each other. If you come to my hone and say you are thirsty I won't give water to you because then what will my family drink. We even watch shacks burn and say there is nothing we can do' (Harvey, quoted in Cock 2007: 63). Another commented: 'Cooking, cleaning and especially washing clothing uses a lot of water. I'm tired of this government. The whites were better. They never cut our water even when we did not pay for it' (Harvey 2003: 5).

While initially residents, burdened by a sense of personal failure in not meeting household needs, retreated into the private sphere of the household, increasingly they are recognizing their personal troubles as a public issue. Whether these local initiatives will become a national movement against essential service privatization remains to be seen. The South Korean experience offers insights into an organized resistance to the privatization process.

The Privatization of Electricity in South Korea

When General Park Chung Hee came to power through a military coup in 1961 he emulated the Japanese model of state capitalism, which harnessed the power of the state to accelerate rapid industrialization through actively supporting the development of large conglomerates (chaebols). Key facets of the economy such as banks were nationalized to increase the power of the state to direct this process. In this context, Park moved swiftly to restructure the production of electricity, which was critical to the new industrialization strategy. Two months after the coup three regional electric companies (Chosun, Kyungsung and Namsun) were merged into a single entity, the Korea Electric Company (KECO), which was a partly state-owned entity. After a further military coup in January 1982 led by General Chun Doo-whan, electricity production was nationalized and KECO was renamed the Korea Electric Power Corporation (KEPCO).

The first sign of a shift in strategy was evident in August 1989 when the government listed KEPCO shares on the Korea Stock Exchange, selling off a 21 per cent stake in the state enterprise. The transition to liberal democracy in 1993 ushered in both a democratically elected government and an embrace of market politics, which included privatization. The newly elected Kim Young-sam government listed KEPCO shares on the New York Stock Exchange and American depository receipts of $300 million were issued in October 1994. In February

1994 Kim published a privatization plan, which aimed at privatizing 58 of the 133 publicly owned companies. The Asian financial crisis of late 1997 increased external pressures on the government to introduce market efficiencies. The IMF and the World Bank stated that loans would only be forthcoming if the privatization plans were realized and labour market flexibility introduced. The Kim Daejung government (1998–2002) decided to privatize the most lucrative state-owned companies, including KEPCO.

Given the history of resistance to military dictatorship, South Korean society has a tradition of militancy. By the 1990s clandestine organizing in the factories had given rise to a vigorous democratic labour movement linked to the student movement and other civic groups. This movement-in-the-making was at the forefront in the campaign against electricity privatization. As a result, the government jettisoned plans for immediate privatization, while still preparing the way for a future move by splitting the utility into six power generation entities in a pre-sale restructuring process, which was presented as a strategy to increase competition.

Following on from a document prepared by the Ministry of Trade and Industry in January 1999, entitled *A Basic Plan for Electricity Restructuring*, the Kim Daejung government continued to prepare for the complete privatization of KEPCO, submitting three bills to the National Assembly in October 1999. The National Assembly did not pass the bills due to pressure from civil society. Opinion polls at the time revealed that 64 per cent of the people were against electricity privatization. However, the government was determined to achieve its goal. The bills were resubmitted with revisions to the Assembly in July 2000 and they were finally passed in the December plenary session. This led to the full privatization of KEPCO in February 2001 and six generation companies (Korea Hydro & Nuclear Power, Korea South-East Power, Korea Midland Power, Korea Western Power, Korea Southern, Korea East-West) were established in March 2001 after KEPCO's general meeting of stockholders approved the split into six companies. In June 2001 the government created a task force, which planned the implementation of privatization.

Organized labour responds

Organized resistance was led by the KEPCO union (Korean National Electrical Workers Union) with a membership of 20,000. It is a Federation of Korean Trade Unions (FKTU) affiliate that was established

as far back as September 1961. As a FKTU affiliate, one would antici-
pate the characteristics of a pliant company-style, state-sponsored
union. However, the depth of civil society feelings against electricity
privatization and their own feelings of insecurity produced by the
restructuring proposals tended to radicalize the membership of the
KEPCO union. The membership from the power generating plants
demanded a more democratic and accountable leadership within the
union, while others split and formed the Korean Power Plant Industry
Union (KPPIU) after KEPCO was split into six power companies.
Significantly, this new union affiliated with the KCTU. In July 2000
the KEPCO union announced a union reform plan to create internal
democracy through the direct election of leadership by rank and file
members through a national election. An election was held and the
existing leadership was returned to office. Internal democracy gave
voice and created participation, which had the potential to lead to a
new movement direction in the union. However, this partial transformation was to have ramifications as the struggle against privatization
unfolded. The immediate effect was to shift the union to a mobilizing
frame as the leadership was forced to respond to demands from the base.
Mass rallies to mobilize against privatization were organized jointly
with other public sector unions and civic groups.

Three major public sector unions, the National Railway Workers
Union, Korea Telecommunication Union and the KEPCO Union, united
forces in a national campaign to block electricity privatization after
the government re-submitted the three electricity privatization bills
to the National Assembly in July 2000. This crystallized into the
Public Sector Union Coalition comprising public sector unions from
both FKTU and KCTU, which was formed on 1 November 2000. As a
further sign of an active civil society, the Popular Coalition for Anti-
Privatization of the Public Sector, which united various civil society
movements, was formed in mid-November 2000. These organizational
initiatives strengthened resistance, which enabled the mobilization of
tens of thousands of public sector workers against electricity privatization in the second half of 2000. Within these developments, the
KEPCO union played a key role.

The state's commitment to electricity privatization, regardless of
mounting opposition, set the stage for confrontation. In a clear demon-
stration of how transformed the KEPCO union was after being democra-
tized, the union decided on strike action, something quite foreign to
the pliant FKTU tradition. In late November 2000 the Public Sector
Union Coalition organized a rally in downtown Seoul: 20,000 workers

participated. Then on 3 December, 4,000 KEPCO union members went on strike. However, management responded swiftly to draw the KEPCO union leadership into a mediation process in the National Labour Relations Commission. Without formal feedback from membership, leadership compromised on the fundamental principle of privatization. In the settlement agreement, the union accepted the privatization of KEPCO and the division of the company into separate units, while the management promised a wage increase. With the deal done, the leadership announced the cancellation of strike action, which was scheduled for the following day. The scene was set for angry confrontation. Thousands of workers gathered at the KEPCO union headquarters and expressed their total opposition to the deal and their anger and dismay that the leadership had not consulted them on the issue. They denounced the leadership, who went into hiding, fearing to meet their members to explain the deal. Having won union agreement, government moved swiftly. The next day, 4 December, the National Assembly Standing Committee on Industry and Trade, who were responsible for the privatization process, passed three bills on electricity privatization.

The compromise of the KEPCO union leadership led to a split in the union at the time when the government was implementing its new law on privatization, which comprised the legal endorsement of the six independent power generation plants. In July workers who were angry over the compromise decided to establish a new union independent of the KEPCO union. They formed the Korea Power Plant Industry Union (KPPIU) with 6,000 members, which affiliated with the KCTU a month later. The KEPCO union retained 13,000 members, with a further 3,000 in the Korea Hydro & Nuclear Power Union. Not surprisingly, in October 2001 management took a hard line with the new KCTU union, refusing to bargain over the union's demands for a new collective agreement granting a wage increase of 12 per cent, the reinstatement of dismissed union activists, the improvement of working conditions, and the complete abandonment of the privatization agenda. Management refused to meet the union leadership and ignored their demands, while the government provocatively released the details of its privatization plan in the document 'Basic Plan for Power Plant Privatization', which stated that two power generating plants were to be privatized during 2002.

At the January 2002 Congress of the KPPIU, delegates decided to strike if these privatizations went ahead. On 25 February the union went on strike and were joined by two other public sector unions (Korea Railway Workers Union and Korea Gas Corporation Union), which were also under threat of privatization and the restructuring and job losses that would stem from this initiative. However, these two unions made an early exit, leaving the KPPIU on its own. The KPPIU strike was a remarkable event. As was to be expected, the state mobilized the full weight of its repressive force against the strike, for these were workers in an essential service from a radical union federation.

In the initial days of the strike 19,500 workers gathered on various university campuses as protected space. However, it soon became clear that the full force of the state was being mobilized against the strikers. Prime Minister Lee Han-dong sounded an ominous warning in a television broadcast: the strike was an illegal, criminal act that must be punished. Warrants of arrest were issued against 47 union organizers and 252 shop stewards and the companies fired these officials, warning that they were 'saboteurs of the very law and order system of the state'. Military police were mobilized to enter university premises. In these circumstances the KCTU made a decision to enter 'uncharted waters' on the night of 26 February, the second day of the strike.

The 5,300 electricity workers divided themselves into small groups of five to ten and left the campuses under cover of night and went to stay in motels in the country towns surrounding Seoul. The KCTU kept them linked through cyberspace. KCTU used the web to generate daily discussion topics and news on developments in the negotiations. Bulletin boards were the means to leave messages. The government responded by attempting to shut down the website and the KCTU threatened similar action on government sites, including that of the Presidential Palace and the National Assembly. The KCTU used a technique of mirror sites, where you set up a website that looks exactly like the one you are targeting, but with messages of protest. This is a clear example of the use of logistical power.

This event received remarkable support from households, who organized support rallies and demonstrations. The commitment of ordinary citizens was strong, with motel owners warning of police presence. Unions from other sectors staged solidarity strikes. These different segments of civil society united in this anti-privatization struggle, determined to support and protect workers in the front line.

The strike ended after 38 days. Its success resided in the fact that it had brought the issue of privatization to the forefront of national politics. The weakness was that it again highlighted the structural divisions in organized labour when the FKTU failed to support the action. In the aftermath of the strike there were further significant developments.

In April 2002 new leadership of the KEPCO union, which was opposed to privatization and which was more willing to form a movement committed to building alliances, was elected. In October they organized a joint mass rally against electricity privatization with the KCTU and FKTU, which was attended by 10,000 workers and citizens.

There was also a significant political development, which altered the landscape. Roh Moo-hyun won the Presidential election of December 2002. In his campaign he had promised to reconsider privatization. In March 2003 he announced that the network industry (railway and transmission/distribution of electricity) would not be privatized, but power plant privatization would proceed. In June 2003 KEPCO was invited to a forum to deal with the restructuring policy on electricity, organized by the Presidential House, and participated in the Special Committee on Public Sector Restructuring of the Korea Tripartite Commission. In the meeting KEPCO called on the Tripartite Commission to adopt a recommendation to stop governments' unilateral policy of splitting the utility into different entities in order to privatize electricity. In August 2003 the Special Committee on Public Sector Restructuring of the Korea Tripartite Commission decided to organize a joint government-labour research team to reconsider privatization. In addition the union organized a Solidarity Meeting in July 2003 where electricity-related unions like Korea Power Industry Union pressurized the government to accept its demands. Finally, President Roh said that his government would accept the result of the joint research team at a meeting with FKTU leaders in September 2003. In June 2004 the Special Committee on Public Sector Restructuring of the Korea Tripartite Commission adopted a resolution to stop the split of distribution parts from KEPCO. The result was that the government's policy for privatization and restructuring of the electricity industry was suspended.

Conclusion

In all three countries there has been a shift towards the privatization of essential services. The most advanced privatization of electricity is in the state of Victoria in Australia, a process which 'hollowed out' democracy and was undertaken behind the backs of citizens. The goal was to reinforce an enterprise culture where individual identity is constructed around consumption to marginalize active citizenship. In South Africa, privatization is most advanced in the provision of water where the world's leading water TNCs have formed partnerships with local companies to provide water to local authorities. Unlike Australia, however, this policy shift has met with considerable resistance in certain local authorities. In Korea the attempt to privatize electricity was successfully blocked by organized labour. In Korean society the citizenry retains a notion that the state should provide essential services.

A state with a strong sense of public interest is a state that asserts the multiple needs of society against the dictates of the market, thereby creating a more secure society where the state intervenes to meet human needs. Privatization heralds the state's withdrawal from this pivotal social role. Once corporations control key areas of human need, social relations are reduced to the status of commodity relations where everything is measured by the market. Nothing is sacrosanct as corporations target public sector activities to which people are attached as citizens rather than consumers. Privatization reflects a captive state that remodels its activities along corporate lines, while reducing its exposure to political pressures from the electorate.

Essentially, the discourse on privatization involves a shift from a public realm view of individuals as citizens to a private realm view of them as consumers (Marquand 1997: 37). This is why role definition is taken so seriously by neoliberalism. Hayek realized that a market order could not be achieved in society where non-market values of active citizenship, cooperation and solidarity prevailed (Hayek 1979). Hence the focus on transforming beliefs, values and self-identity, placing cultural reconstruction at the heart of the neoliberal project. Notions of active citizenship, solidarity cultures and the public interest have to be transformed into those of an enterprise culture where the consumer reigns supreme and where entrepreneurial values, subject to market requirements, dominate. This is a concept where the state chooses to ignore social class and inequality by assuming that where a need exists, there are consumers who are able to pay for it. It is for this reason that neoliberalism asserts a hedonistic definition of the self that challenges notions of active citizenship.

In Part One we have examined the impact of markets against society. We now turn to an analysis of the responses of society in the three towns to the market.

Part II

Society Against Markets

In Part One we showed how workers in all three countries are experiencing a growing sense of insecurity as the employment relationship is reconfigured, through the introduction of short-term contracts, outsourcing or the threat of competition from 'the China price'.[1] We showed how liberalization has led to corporate restructuring on a global scale resulting in growing concentration and centralization. We also analysed the growing commodification of essential services through privatization.

To understand increasing insecurity in the workplace, it is not sufficient to identify the heightened power of global corporations and the convergence of labour processes and management practices, as Nichols and Cam (2005: 208) in their study of the white goods industry do. Instead, there is a need to be context and place specific, if one is to understand the impact of globalization on labour and go beyond the workplace to examine the household and the community. In Part Two we enter the 'hidden abode of reproduction' and the communities in which these households are located, to examine the impact of restructuring on workers' lives. We show how households in Orange, Changwon and Ezakheni are structured differently and respond differently to these external pressures as they search for security.

The household is a site where incomes are pooled and a reduction in wage earning places pressure on this vital social institution, forcing members to search for other sources of income. In Orange most interviewees respond fatalistically and seem to rely on the declining provisions of the welfare state, although there are signs of an imaginative attempt to globalize the trade union. In Changwon, where there is limited social security, especially for irregular workers, and few opportunities for non-wage income, workers respond by working harder through increasing the amount of overtime they do. Households in Ezakheni are more 'flexible' and are able to compensate for a loss of wage income by engaging in a range of non-wage income activities. However, these non-wage income activities are not sustainable and we identify a growing crisis of social reproduction in this rural town.

We examine first the changing employment relationship and how workers in these three countries are responding to these pressures. Two broad responses are identified in each site. The first is that of retreat, which can take the form of a fatalistic acceptance of the changes, retreat into the household as a survival strategy, or an adaptation to the market by working harder, or, more dramatically, migrating to another part of the world or country.[2] The second is an attempt to resist through collective action. The first reaction tends to be an individual response to what we referred to earlier as 'personal troubles', but it can also be a collective response, such as company or business unionism, a collective acceptance of existing relations of market power. The second response is an attempt to turn a personal trouble into a public issue by engaging in collective action to challenge the power of the market, the corporation or the state. However, a challenge of growing insecurity can lead to an individual response, such as industrial sabotage, or people asserting their rights as individual citizens.

A typology of responses to insecurity

	Individual	*Collective*
Acceptance	Work harder – market adaptation Retreat into the household, fatalism Migration/emigration	Company or business unionism
Challenge	Industrial sabotage Active citizen	Collective mobilization and a search for alternatives

The important point is that 'the economic position of workers is not necessarily determined by individual characteristics of that worker, be it age, gender, or employment status … a workers' ability to survive, and her economic position, is dependent not only on herself, but on a whole household full of people, notwithstanding the fact that households may not allocate resources equitably' (Kenny 2001: 102). Indeed, there is a qualitative difference in households which have access to a 'good job' – a formal sector, secure job – and households that do not. Those who do not are not only poorer, but they tend to have fewer resources to assist them in other non-wage means of provisioning (Nelson & Smith 1999).

How, then, do we conceptualize the impact of the reconfiguration of the employment relationship on the structure and composition of the household? The household is an income-pooling unit. It is 'the social unit that effectively over long periods of time enables individuals, of varying ages of both sexes, to pool income coming from multiple sources in order to ensure their individual and collective reproduction and well-being' (Wallerstein & Smith 1992: 13). Wallerstein and Smith identify five major forms of income: wages, market sales (or profit), rent, transfer (such as state grants, remittances or inheritance) and 'subsistence' (or direct labour input) (Wallerstein & Smith 1992: 7–12).The changing nature of the household, they suggest, can best be understood as a response to cyclical patterns in the expansion and contraction of the global economy:

> Global contraction will lead to squeezes which force units of production to find ways of reducing costs. One such way of course is to reduce the cost of labour. This may in turn lead to changes in the mode of remunerating labor ... A household is a unit that pools income for purposes of reproduction. If the income it receives is reduced, it must either live on less income or find substitute income. Of course, there comes a point where it cannot survive on less income (or survive very long) and therefore the only alternative is to find substitute income. (Wallerstein & Smith 1992: 15)

This process of expansion and contraction and its impact on the household is exacerbated by the continuous restructuring of work, even in the expansionary phases of capitalism. Households that are least able to 'find substitute income', they suggest, are those most dependent on wage income. Those households which can most readily invest in non-wage activities are the most 'flexible' as they can increase their income by autonomously engaging in such non-wage activities as subsistence activities, renting out a room, or trying to secure additional transfer income from the state. But the ability to secure non-wage income is itself a function of the boundaries of the household. A small family may not have the hours available to generate the necessary non-wage income. As a result changes in the world-economy create pressures on household structures that either expand their boundaries or shrink them, depending on their needs and resources (Wallerstein & Smith 1992: 15–16). This conceptual framework is useful in making sense of the differences in the nature of the households in the three sites.

These varied potential household responses to corporate restructuring are rational attempts to respond to changed material circumstances. However, as we explore in Part Three, social movement unionism could complement these *individualized* household responses by organizing and mobilizing both at the workplace *and* households in the community. In the final part of the book, we will draw on the experience of certain unions in the South which have adopted this strategy.

An obvious point to make is that the sizes of the households in the three research sites are very different. This suggests intriguing differences in the ways in which these households search for security. What is clear from our comparative examination of households in Australia, Korea and South Africa is that these core institutions are always embedded in a specific social context. It is to this examination we now turn, beginning with Ezakheni, South Africa.

5

Strong Winds in Ezakheni

The Place ...

Ezakheni/Ladysmith is the home of Ladysmith Black Mambazo. This African male vocal group became known beyond South Africa's borders after they featured prominently on Paul Simon's album *Graceland*. In the song 'Homeless', they sing about the landscape around their home village. They sing about 'strong winds' that destroyed their homes and about the many people who died in the area. The lyric reminds us of the trauma of apartheid's forced removals and of civil war. The chorus is hauntingly beautiful:

> And we are homeless, homeless
> Moonlight sleeping on a midnight lake ...[1]

Ladysmith was a typical South African colonial settlement. The town attained its name from the Spanish wife of a colonial administrator Sir Harry Smith – Juana Maria de los Dolores de Leon, or known to the locals as Lady Juana Smith. It was formally proclaimed as a township in 1850 by the British colonial administration of the Colony of Natal and is located in one of the most spectacular of South African landscapes. The Tugela River flows nearby, and to the west is the Drakensberg mountain range, which is often covered in snow during winter. The town itself is surrounded by hills covered in African savannah and acacia trees. Today, Ladysmith is marketed as a tourist haven. Nevertheless, despite the natural beauty of the surrounding landscape and a number of quaint Victorian buildings, the strong winds of history have left many homeless in and around this town.

Local authorities and tourist companies in the area prominently market historical battles that took place here as the Battlefields Route. Some of the battles in this part of the country shaped the course of South Africa's history. These include the Battle of Blood River, where the Boers defeated the Zulus in 1838, the Battle of Isandhlwana, where the Zulus defeated the British in 1879, as well as the Battle of Amajuba, where the Boers defeated the British in 1881. A somewhat nostalgic cottage industry has developed around these battles, and enthusiasts of colonial history annually re-enact events.

Ladysmith itself was the site of a spectacular siege. In 1886 a railway line was established between the gold fields of the Witwatersrand and the Durban harbour, with Ladysmith as one of the stopovers. This led to local economic growth, but also war. The discovery of gold on the Witwatersrand and the resulting tussle between the Boers and the British Empire led to the Second Anglo-Boer War which began in October 1899. Because of Ladysmith's logistical position on the transport route, Boer forces besieged the town for 118 days between November 1899 and February 1900. The end of the siege was celebrated as far as Orange in Australia, where the local racing club referred to the incident in an invitation to local residents to attend the races in the local newspaper. In 1905 a monument was unveiled in Orange to commemorate the Boer War. Ladysmith's name appears on the monument, along with Mafikeng and Kimberley, two other towns that were under siege, and Pretoria, where the peace agreement was signed to end the war (Nicholls 2005: 27, 99–100). But very few present-day residents of Orange are even aware of this memorial. In Ladysmith, however, the siege is still celebrated in the local museum.

All of these celebrated military battles happened in the 1800s. The more recent strong winds of history – battles against apartheid, spatial planning and its forced land removals, as well as the civil war between the ANC and Inkatha, a Zulu ethno-nationalist grouping – are not celebrated or commemorated publicly. While the local library has ample books on the Zulu War and the Anglo-Boer Wars, we tried in vain to find accounts of forced removals and the civil war of the late 1980s and the early 1990s. This is where Ezakheni enters the story.

The authorities under apartheid, drawing on Africans forcibly removed from Matiwanes's Kop and Lime Hill, created Ezakheni to serve as a labour reserve to implement the apartheid state's post-1970s policy of industrial decentralization. This was intended to stem the urbanization of Africans. In the 1980s this strategy had the additional function of circumventing democratic unions, since factories in 'homelands' were excluded from South African labour law reforms which gave independent unions the space to organize. The industrial park in Ezakheni became one of the flagships of this policy. Both the park and the residential area were located within the borders of 'trust land' – or KwaZulu, the Zulu 'homeland', which was administrated by Inkatha. The area had the added advantage of propinquity, since it is a mere 15 kilometres from Ladysmith, where white and Asian factory owners, managers and artisans resided.

The end of apartheid also meant the end of subsidies from the central state to keep these decentralization zones functioning. Many factories closed down, but in certain parts local governments stepped in to maintain some form of subsidized services. Unlike many of the textile and clothing firms that closed shop in Ezakheni, Defy is still operating and is even expanding. Despite this, and because of some of the firm's employment practices, workers have a profound sense of insecurity. Furthermore, there are very high rates of unemployment in Ezakheni, and many family members depend on the wages of Defy workers. Indeed, the real differences between Ezakheni, Changwon and Orange emerge when one enters the hidden abode of reproduction, which we attempt to do in this chapter. But first, we consider how workers respond to their feelings of insecurity in the workplace.

Employment and Insecurity

In Ezakheni all the full-time workers we interviewed felt that their employment is secure, except for one respondent who said: 'I do not feel very secure particularly when I look around and see all the other cheap appliances, such as those imported from China. And I feel scared that because of lack of money people will opt for cheaper brands and we may lose our customers.'[2] The sense of security among full-time workers can most probably be explained by the fact that Defy is seen as one of the more stable firms in Ezakheni.

In sharp contrast, short-term contract workers (or STCs, as they are colloquially referred to) felt insecure. Five said that felt very insecure; as one remarked: 'I am not secure at all. I don't know when I can be laid off.'[3] In a similar uncertain tone, an STC worker observed: 'You never know what is going to happen. It is the same with overtime; you are only told in the morning that you'll be doing overtime.'[4] This was confirmed by a further worker: 'My contract can be terminated any day. I cannot even open credit accounts ... because I may not have enough money to pay my instalments.'[5] Finally, 'I don't feel secure since I am a contract [*sic*]. I cannot even apply for any of the internal posts, because us contracts are taken as though we don't exist. We are only given a 1–2 weeks' notice if we are going to be laid off.'[6]

Still, compared to the many residents of Ezakheni who are unemployed – an estimated 49.2 per cent (Emnambithi Local Government 2002: 4) – contract workers consider themselves lucky. Two of the STC workers we interviewed were ambivalent. One remarked: 'Yes, because when they need you, you are always called. They will always consider you first. At the company we are treated the same as the permanents. My only problem is I don't know when I will be laid off.'[7] Another said: 'One can never be 100 per cent sure of employment security. As long as there is no competition from the Chinese, I feel secure.'[8] To be sure, two of the STC workers – both of whom had accumulated some levels of work experience and skill – felt relatively secure. As one remarked: 'I feel fairly secure as compared to the job I was previously doing as a stock controller at Dunlop.' Another STC worker, anticipating a change in status, said: 'I feel fairly secure because I know that my permanent registration is near.'[9]

Having a contract job in Ezakheni is considered to be better than having no job at all, especially where just under half of the economically active population are unemployed. Clearly, it is a case of with whom you compare yourself, a process of relative deprivation, where your sense of deprivation is lessened when you compare yourself to those who are in a less fortunate position. The sense of relative security is premised on the fact that others are even less secure, for the segmentation of the labour market creates a hierarchy of insecurity. It also creates a sense of relativity and, in doing so, division among workers.

We asked respondents whether employment security had changed over the past five years. In Ezakheni three STC workers felt their sense of security had changed. Others felt that the more years they worked, the more secure they were.[10] Because their registration as 'permanents' was imminent, they felt more secure. The other six contract workers felt that their insecurity had remained at the same level. However, those workers who had become permanent had an increased sense of security since they had experienced the benefits of certain rights in the workplace[11] and they have acquired more skills.[12] While another worker, who did not feel secure, said, 'I've always felt that my job security was on and off. Recently, I do not feel very secure.'[13]

In Ezakheni most of the STC workers were actively adapting to the market by either undergoing additional training or applying for more stable jobs either in the firm or elsewhere. For example, one worker was studying 'criminal justice' to secure a job in the Department of Correctional Services.[14] Another had completed a sewing course[15] and a further contract worker was doing a pre-university course with the University of KwaZulu-Natal in order to qualify for acceptance in a course in engineering.[16] Four had applied for other, more stable jobs, all in government departments.[17] The permanent worker who felt insecure said, 'I have tried to study. I once enrolled in a computer course, but I dropped out after a few months because of the shift system which did not allow me sufficient time to do the practical work.'[18]

Others displayed a fatalistic attitude towards changing employment conditions. For example, one respondent remarked, 'In this firm the environment doesn't encourage one much. In fact I am tired of this firm's atmosphere. Right now I am here mainly so I can sleep on a full stomach. I would leave the firm without thinking twice if another form of employment arose.'[19] Five of the other permanent workers felt that there was 'not much' or nothing else that they could do to improve the security of their jobs.

Four of the STC workers felt that if they became permanent they would have a much better chance of improving their security by applying for a supervisor's position,[20] going for training,[21] getting a bursary to study at college.[22] One worker said, 'I think I need to put more effort into the work that I do and be more dedicated. Maybe I might be lucky to get registered [i.e. get a permanent position].'[23] The other four were fatalistic about their chances of improving their security. One worker said, 'Even if you were to work harder it is all in vain; nobody notices. Studying is even worse; it does not benefit you within the firm.'[24] Another pointed out: 'I've heard other workers say studying does not help in this firm.'[25]

Households and Insecurity

Compared to the households in Orange and Changwon, those in Ezakheni tended to be very large, consisting not only of a nuclear family, but also of a range of dependents and extended family members. For those we interviewed, the average household size was seven. This is above the average household size in KwaZulu-Natal, which is 4.8 (Statistics South Africa 2004).

The structure and function of households in Ezakheni differ substantially from those in Changwon and, in particular, Orange. It is clear from our research that most of the households in our Ezakheni sample do not receive a living wage. As a result, members of these households have increasingly had to find work outside wage labour in order to supplement their wage income.[26] Importantly, they are not a privileged sector cut off from society, but are rather deeply embedded in communities of poverty, sharing their earnings among their large households. Low wages thus affect everyone in Ezakheni.

At the core of our research findings is a systematic comparison of the two groups of workers we interviewed in Defy: those on short-term contracts (STCs) and those who are full-time workers (Fakier 2005: 36–47). A systematic comparison of wage levels, as well as household income and composition, shows considerable differences between permanent employees and STCs. In terms of actual wages, permanent employees generally earn nearly double the wages of STCs.[27] One also has to consider that the overall household income levels of the permanent workers we interviewed tended to be much higher than those of STCs – in fact, these levels are two thirds higher. There also seems to be a strategy of clustering, where households with higher incomes tend to attract more members as a coping strategy (see Mosoetsa 2005). Indeed, the STC workers tended to have fewer children in their households – mostly one child per household – whereas permanent employees we interviewed often had three children in their households (see Fakier 2005).

As table 5.1 demonstrates, these STC workers cannot survive on their wage incomes alone. They are forced to engage in non-wage income generating activities such as selling clothing and shoes, undertaking electrical repairs, selling linen, driving a brother's taxi and plumbing. Of course, they also obtain income from other household members, as well as state transfers, such as pension grants, but the bulk of their non-wage income is derived from what we described earlier as non-wage 'market' income activities. Indeed, the highest earning STC worker at Defy draws nearly as much income from her non-wage 'market' income activities every month (R2,000) as she does from her employment at Defy (R2,233,33).[28] In contrast to STC workers we interviewed, the full-time workers tended to get very little income from non-wage activities. Indeed, only one full-time worker gets any non-wage 'market' income – he estimates this income to be around R1,000 a month for his activities as a herbalist.

Unlike in Orange where household responsibilities do not go beyond the small nuclear family, in Ezakheni the household includes an obligation to pay for more distant relatives, such as a nephew's tertiary education. Furthermore, as Murray demonstrates in his study of migrant

Table 5.1 Ezakheni, income and expenses of STC workers, 2005

Number of STC workers interviewed	STC wage income when working at Defy	Total income of household including income from other members and sources	Total expenses	Total income – total expense	Gap between total household expenses and STC wage income
3	2,680.00	5,973.33	4,080.00	1,893.33	−1,400.00
6	2,700.80	3,360.53	2,405.00	955.53	295.80
7	2,700.80	3,080.53	3,669.00	−588.47	−968.20
9	2,720.00	2,720.00	2,600.00	120.00	120.00
10	2,800.00	3,553.33	2,750.00	803.33	50.00
11	2,700.00	3,050.00	2,875.00	175.00	−175.00
12	2,480.00	1,653.33	2,170.00	−516.67	310.00
16	2,700.80	1,575.47	2,655.00	−1,079.53	45.80
18	2,700.00	1,625.00	2,370.00	−745.00	330.00
Averages	2,686.93	2,954.61	2,841.56	113.06	−154.62

Source: Fakier (2005: 46)

labour in Lesotho, households in Southern Africa are not a co-residential group; 'the energies and resources are divided between a variety of activities in Lesotho and South Africa' (Murray 1981: 47–48). Money and other resources are distributed beyond the boundaries of one place of residence, increasing the household size significantly.

Ironically, STC workers in Ezakheni save more on average than full-time workers. Among those STC workers we interviewed, an average of R875 was saved monthly, while the full-time workers saved an average of R513. Clearly, STC workers are more insecure and try to increase their savings to help them survive when they are unemployed. Indeed, three STC workers said that, while their wages were not enough to sustain their basic needs, they were able to survive by 'limiting' themselves.[29] Or, in another example, the respondent remarked, 'I manage fairly well, but when I am laid off it is difficult and I rely on my plumbing job.'[30]

Responses of Defy workers in Ezakheni suggest that there are two levels of insecurity, two different modal or ideal-type responses. On the one hand, there are those in full-time work who feel a degree of security and belong to the trade union, NUMSA, and engage in less non-wage income activities. On the other hand, there are those on short-term contracts who constantly feel insecure, feel they have to work harder, do not belong to NUMSA, and are constantly searching for new employment opportunities both in Ezakheni/Ladysmith and by migrating to other urban centres (Fakier 2007). KwaZulu Natal has the third highest rate of migration (after Eastern Cape and Limpopo) of the nine provinces; 78,444 people migrated between 1996 and 2001, with this figure steadily increasing in the past five years (PCAS 2006: 55). These responses confirm our identification in chapter 8 of three zones of work: the core, the non-core and the periphery.

However, what distinguishes households in Ezakheni from those in Orange and Changwon is that Ezakheni is surrounded by the legacies of the traditional homestead economy of KwaZulu-Natal. This is reflected in the households of 2006. Drawing on Polanyi's concepts of reciprocity and non-market relations introduced in chapter 1, we would suggest that a culture of sharing remains in these households, as does the large extended family where children are expected to share in the work tasks of the homestead. Household members engage in multiple economic activities, which include the herbalist drawing on indigenous knowledge or others making and selling commodities.

However, subsistence (in the traditional sense of being separate from the market economy) has ceased to be an alternative for these households. Furthermore, it is misleading to see the non-wage income activities of Defy workers as a 'second economy'. 'Most of the economic activities in the periphery are dependent on markets created by formal economy activities. It is likely that such activities will be able to expand only to the extent that the formal economy itself expands' (Webster & Von Holdt 2005: 36).

While drawing on non-wage forms of income may help to explain why workers in Ezakheni are able to compensate for low real wages, this cannot be the solution to the STC workers at Defy's search for security. As competition intensifies in the white goods industry globally, Defy managers are 'resolving the crisis of the post-apartheid workplace disorder by displacing confrontation, antagonism and disorder into the family, the household and the community. This generates a broader social crisis whose symptoms are the breakdown of social solidarity, intra-household and community conflict, substance abuse, domestic violence, and the proliferation of other crimes' (Webster & Von Holdt 2005: 31).

This 'crisis of social reproduction' cannot be properly understood if the household is defined simply as an economic unit that pools income. Households are also, as Mosoetsa suggests, a 'social core of socialization, emotional support, caring and feeding of household members' (Mosoetsa 2005: 3). To understand the household 'socio-logically' requires an examination of the intra-household dynamics through ethnographic research. By undertaking such an examination in KwaZulu-Natal, Mosoetsa confirms our finding that the household is emerging as a site of 'fragile stability' in response to the social crisis generated by unemployment and fuelled by HIV/AIDS (Mosoetsa 2005). The household, she suggests, has become a place to which people retreat. It is the major site for sharing economic resources such as housing and income through state grants such as old-age pensions, child and disability grants and grants to those who have HIV/AIDS.[31] Households have become, she suggests, sites of production and reproduction attracting poorer family members in search of security.

However, Mosoetsa is able to demonstrate through her ethnographic research that these households are not homogeneous, tension-free institutions. Sharp conflicts, based on gender and generation, emerge around the allocation of household resources (Mosoetsa 2003: 7–8). Interviewees, Mosoetsa observes, often cited the loss of income through alcohol abuse by unemployed men in the household. Young women who receive child support grants on behalf of their children are accused of spending the grants for their benefit only, especially on cell phones (mobiles), clothes and hairstyles. The power struggles that surround the allocation of resources threaten the potential benefits of these networks in reducing individual and household insecurities. They also lead to high levels of interpersonal and domestic violence.

Households have become places to hide one's poverty and, through links with rural households, places to 'hide away' those with AIDS (Mosoetsa 2003: 8). With declining household incomes, members are no longer able to make monetary contributions to stokvels[32] and burial societies and are instead offering 'in-kind help such as cooking and baking and lending the bereaved family dishes and pots during funerals' (Mosoetsa 2003: 12). The nature of such relationships is based on reciprocity and those 'households that are known for helping others get more support from the community than those who do not' (Mosoetsa 2003: 12).

Restructuring, Political Parties and Community

In Ezakheni, the ANC is the hegemonic political force and 14 of the 19 workers we interviewed were members of the ANC. However, there is growing disillusionment with the capacity of the ANC to deal with the pressing issues facing the community. Five of the respondents did not belong to the ANC and were not members of an alternative party. One respondent explained why he no longer had any faith in political parties, 'because these parties want membership by promising us many things. I once joined the IFP[33] because it was the leading party here and they promised to help us with bursaries and work. But all we did was to go to rallies.'[34]

The reasons the respondents gave for supporting the ANC were essentially to do with their basic needs. Six said the ANC had given them housing; four said they had been given access to clean water and electricity. Two said they had been given help by the ANC in a community dispute, while another mentioned help for AIDS orphans. As one respondent remarked: 'I have a site where I've built a house for my family which would have taken longer time to obtain [if it was not for the ANC]. I can have a say in my community if I wish to.'[35] However, a female contract worker who was a nominal member of the ANC was more sceptical in her support: 'I belong to the ANC in name only. I'm not an activist ... I have been promised job opportunities but all I have is this contract job.'[36]

The roots of the ANC lie deep in these areas and their present responses are shaped by their broader commitment to the ANC's role in the national liberation struggle. As one observed: 'The ANC is the organization that I have grown under and many people sacrificed their lives for it. One such person is my brother who died during the 1991–1993 political violence.'[37] Given the chance, I believe the ANC can do wonders.'[38] Another respondent introduced an important qualification by saying that he would only support the ANC if it was led by its controversial deputy president, Jacob Zuma: 'No, except if its leader was Jacob Zuma. I love the ANC and have great faith in its leaders. Particularly I love Zuma because he seems to understand the working class.'[39] A 31-year-old male contract worker commented, 'I don't think there will ever be a more stable party like the ANC. Some come up and disappear. It is one of the more credible organizations.'[40]

Four respondents made it clear that they no longer had an interest in party politics. One said: 'Political parties do a lot more talking than actually acting. At least I am used to the ANC and its shortcomings. Anyway it would not make much of a difference because I am not active anyway.'[41] Another, a 51-year-old male, permanent employee at Defy, who was a Seventh-day Adventist, said: 'My religion does not allow me to belong to a political party, but I align myself to the ANC.' When asked if he would join a political party to improve conditions at work, he said, 'I am tired of politics'.[42] Another permanent employee, a 26-year-old female, said: 'Politics do not interest me. I just live my life quietly day by day. Most of the time politicians just talk to get votes; they are not really dedicated to the community.'[43] A younger female contract worker (21 years old) said: 'Politics is not my thing. I am not interested in them with their empty promises.'[44] Only one respondent suggested the need for an alternative political party: 'It would have to be a new party, not any of the existing ones, but first I would have to see it keeping promises.'[45]

To understand the hegemonic position of the ANC in Ladysmith/ Ezakheni it is necessary to delve deeper into local resistance to racialized dispossession. As Gill Hart has shown, the Ladysmith/Ezakheni area was one in which significant opposition to the apartheid state's attempts to dispossess black people of their freehold land (the so-called black spots) took place in the 1960s, 1970s and 1980s (Hart 2002: 96–126). Indeed, the legendary ANC leader Govan Mbeki led a 'less elitist form of ANC-organizing' in the area in the early 1950s and many of those who were removed to Ezakheni and the surrounding areas in the 1970s were part of a tradition of local resistance (Hart 2002: 97–98, 120–126). When industry was established in the area, union activity followed in the late 1970s and early 1980s, led by what was later to become

NUMSA (Hart 2002: 109–110, 123–126). As Hart writes: 'The over-lapping land and labour movements in and around Ladysmith represent a locally specific form of social movement unionism that flies in the face of claims that the articulation of workplace and community politics was a distinctly urban phenomenon' (Hart 2002: 110). This alliance included youth activists, as this observation by a unionist who had been closely engaged with support of the youth movement illustrates: 'Many people here are still in the places where they were born. Our resistance to forced removals brought us together' (cited in Hart 2002: 125).

This anti-apartheid alliance that emerged in Ladysmith and its environs in the 1980s wove together, Hart suggests, 'Zulu patriarchal sentiments and practices associated with an agrarian past merged with those that were startlingly new' (Hart 2002: 125). It was also to lead to strong support for the ANC when the first local government elections took place in 1996. In Ladysmith the ANC won 62 per cent of the seats, roundly defeating the IFP which emerged with only one seat (Hart 2002: 241). The result was that the town clerk of Ladysmith resigned, along with at least six other senior municipal officials including the borough engineer, the town treasurer, and the chief of health services (Hart 2002: 244). When the new, democratically elected ANC councillors took over in February 1996, they immediately convened 'large and extremely lively open-budget meetings at which residents were explicitly invited to educate the councillors about their priorities' (Hart 2002: 254).

The result of raising the expectations of their constituents was, of course, to confront the 'disabling' impact of globalization. In the latter part of 1996 the newly elected ANC mayor visited India and China as part of a delegation organized by the KwaZulu Marketing Initiative (KMI) to solicit foreign investment. At the end of 1997 a local economic development forum was formed with representatives from capital, labour, the local state and NGOs. This initiative came to a standstill when the unions made it clear that they strongly opposed the foreign investment strategy pursued by the council (Hart 2002: 273–274). That these visits to solicit investment from East Asia continue, and that they circumvent the involvement of the community and the local NUMSA branch, was confirmed by the manager of local economic development in Ladysmith/Ezakheni[46] and the NUMSA local organizer.[47]

While the ANC in Ezakheni managed to attract nearly 70 per cent of the vote in the next local elections in 2000, the open budget meetings they had initiated had led residents to develop a 'sense of themselves as political actors in relationship to elected officials as well as local bureaucrats' (Hart 2002: 285). Importantly, Hart suggests that township dwellers were pressing demands for urban services in the language of rural rights, and invoking histories of forced removals to drive home their claims. These histories went far beyond personal loss and hardship to address much broader issues of dispossession. A number of speakers drew attention to the much lower cost of living in surrounding rural areas that had managed to resist dispossession, contrasting how their own lives had become commoditized, and stressing how any increase in service charges would be an impossible burden. (Hart 2002: 285–286)

In the late 1990s unions had spoken about developing a strategy around the social wage by forging connections between townships and surrounding rural areas (Hart 2002: 287). However, ongoing job losses with large numbers of retrenched workers had begun to impact on the surrounding rural communities, making it difficult for the ANC to win local support. The result was growing realization of the limitations of political parties and a search for security within the community rather than political parties.

The inability of political parties to deal with the magnitude of the social crisis facing the community has led to religious organizations playing a central role in 'healing, community building and teaching' (Mosoetsa 2003: 11). As one worker remarked: 'My religion is more meaningful to me and I see the need to be of more service.'[48] Eight of the respondents said that their community activities were related to their church. Indeed, it was suggested by one community leader that without the support of the churches, government-funded initiatives such as Khomanani[49] would not succeed.[50] Older women in the churches play an important role in house visits to care and pray for the sick, and to offer comfort and support for the bereaved families after funerals. In addition to the established denominations such as Anglican, Catholic and Presbyterian, as well as the traditional African churches such as the Zion church, new religious organizations have emerged, such as the Philippian Youth Revival.

Workers seem to have less time on their hands to engage in such traditional leisure-time activities as soccer. A 32-year-old male worker said: 'I no longer play soccer since I spend more time at work or watching it on TV. Though during holidays I play. I now get tired easily because I often work late.'[51] Another said: 'If I am not working, I am driving the taxi. So now I don't have much time for anything else.'[52] A 24-year-old female contract worker remarked: 'I have lost interest in church and I need my weekends to rest and do my other chores.'[53] Age was also a factor. A 40-year-old male permanent employee says he has

Table 5.2   Community based organizations in Ezakheni

Community based organizations	No. of workers who listed organization
Home-based care	13
Community garden schemes	12
Funeral society	7
Choral society	3
Youth organization	3
Community policing	2
ANC small business development	1
Cooperative	1
Football	1
Malaria prevention	1
Sewing club	1

Source: Fakier (2005: table 14, community activities)

stopped playing soccer: 'Time – I don't have leisure time available. Age also does not make me have the energy I used to have.'[54] Another states that he has stopped playing soccer and no longer goes to church as frequently.[55] This was echoed by a 40-year-old male permanent employee: 'I am no longer a frequent churchgoer. My wife does all the churchgoing together with the children so there is no need for me to go.'[56] In contrast, one worker – a 26-year-old female contract worker – has increased her activities: 'as I grow older I am faced with more responsibilities as the breadwinner in the family. I believe my family relies on me for almost everything. So I need strong support of friends and church during my time of need.'[57]

At the core of the community's response to the social crisis caused by high levels of unemployment, disease and crime, is the emergence of semi-formal community based organizations (CBOs). In a pioneering study, it was estimated that 53 per cent of the non-profit organizations (NPOs) fit into this category. The study suggests that 'informal, community based networks are on the rise, particularly in the struggle to deal with the ever-increasing repercussions of the government's failure to address HIV/AIDS and unemployment crises' (Swilling & Russell 2002: viii). Table 5.2 lists the types of community based organizations and the number of workers who mentioned them.

Twelve of the respondents said they participated in one or more of these CBOs. Many of these activities are coordinated by the CBO network, which is organized around five objectives:

- Social mobilization
- Health, and AIDS treatment in particular, which is implemented through home-based care
- Food security
- Youth development
- Business development[58]

Community gardens, for example, are designed to meet the objective of food security through providing nourishment for the poor and the sick. It also creates jobs for the aged, youth and the handicapped who are employed in these gardens. Crèches have been established through donations. Small businesses have emerged where skills such as beadwork and sewing are utilized to provide employment for the youth.[59]

Many of the activities of these CBOs are gendered and are seen as 'women's work'. 'I do not really have time. It's more of a woman's thing and I don't enjoy activities which involves a lot of people', said a Defy worker.[60] Another said, 'Yes, [I participate in the home-based care organization] if I am available. I help my wife [who cares for their seven children and works] who is a volunteer to feed the people.'[61] Another said, 'I support funerals of members by lending my hands and through financial contributions. I attend all the abstinence meetings and activities [of the youth organization that encourages young people to abstain from sex] because my younger sister is a member.'[62] Yet another said, 'I support mostly the funeral support organization. This is the one that is very busy, because of the high death rate in the community due to illness and crime.'[63] The high death rate can probably be ascribed to AIDS. We then asked respondents to identify problems in the community (table 5.3).

South Africa has the highest murder rate in the world. This is reflected in Ezakheni, where murders have averaged in the past decade over 58 per year, slightly above the national figure (CIAC 2005). Although murders have declined from a high point in 1994 of 87, property crimes (robbery, etc.) have increased. This has led to growing insecurity among the residents of Ezakheni. A community leader in Ezakheni remarked that this is because the more fortunate (or secure) people living in Ezakheni are not involved in community activities to protect those less fortunate. 'If the privileged don't look out for the poor, the poor will disturb them.'[64]

While workers listed crime and drug abuse or alcoholism as the most serious community problems, they saw unemployment as the root cause of these social problems: 'The lack of job opportunities results in people

Table 5.3 Community problems identified in Ezakheni

Problem	Workers who listed problem
Crime	16
Drugs and alcoholism	12
Unemployment	10
Disease	8
Basic services (water and electricity)	6
Poverty	5
Roads	3
Childcare facilities	2
Health support	2
Police inefficiency	2
Lack of recreational facilities	1
Overcrowded neighbourhood	1
Prostitution	1
Transport	1

Source: Fakier (2005: table 15:52)

willing to do anything to make a living, e.g. young girls becoming prostitutes and drug and substance abuse used to drown people's sorrows'.[65] But in the eyes of a community worker the problem goes beyond unemployment to the foundations of society, the youth. 'The youth in primary school sell their bodies to get nice things: to taxi drivers, business men and truck drivers on the N2 road.'[66]

While these grassroots networks provide crucial support in the neighbourhood, they are not seen as having the capacity to solve community problems. A solution, table 5.4 suggests, requires substantial resources and legitimate authority through intervention by the police, local government and business. The role of the grassroots networks, a community leader observed, is to 'make sure that the job of the government is "functioned" – is done. Government at the top is functioning but as it trickles down to the regions and provinces, less happens. We have to take our political hats off and cook one pot of development.'[67]

Fifteen of the respondents said that the trade union organizes activities in their communities. However, they mentioned marches and demonstrations, or sporting activities, rather than community development. The impression given by those interviewed is that the trade union is there to defend jobs and does not play a central leadership role in the community. As one STC Defy worker suggested, 'The trade union

Table 5.4 Who interviewees think should take responsibility to address social problems

Police	18
Local government	18
Defy	13
Community based organizations	7
Trade unions	5
Churches	3

should ensure that jobs are created and fight for people to be registered permanently, at least, after 6 months.'[68]

The demand for 'decent work' brings to the fore the central contradiction facing the local state in the era of neoliberal globalization. This is captured succinctly by Hart:

> In the name of both democracy and efficiency, local councillors and bureaucrats have been called upon to confront massive redistributive pressures with minimal resources. Simultaneously they have been assigned major responsibility for securing the conditions of accumulation under the aegis of 'local economic development'. The local state, in short, has become a key site of contradictions in the neoliberal post-apartheid order. (Hart 2002: 7)

These contradictions can be seen in the policy of the local state towards incentives offered to foreign investors. One of the major incentives for investing in the Ezakheni/Ladysmith area is the abundant supply of water from the Tugela River and the Drakensberg mountains. To attract investors the council has reduced the cost of water by 13 per cent less than that charged to the local community. To pay for this increase, the cost of water to households has been raised from R1.53 to R4.50 per kilolitre, evoking great outrage from the community.[69] Furthermore, water has emerged at the centre of the power struggle between the IFP, which controls the district council in charge of water delivery, and the ANC, which is in charge of the local town council.[70] What is of particular concern to local residents is that the community gardens in Ezakheni do not benefit from the subsidization of water given to investors. In fact, the poor in the local community are subsidizing investors, sharply foregrounding the contradiction facing the local state in post-apartheid South Africa. Not surprisingly, the privatization of municipal services such as water, as described in chapter 4, has been the most overt source of conflict between unions and community organizations on the one hand, and municipal authorities on the other, often placing township councillors in a difficult position.

Our analysis of the role of the less formal community based organizations in Ezakheni is that, in alliance with the trade unions and the local ANC, they could provide the basis for an alternative local economic development strategy to that of the current neoliberal orthodoxy.

Conclusion

This account of household and community responses at Ezakheni illustrates a fundamental contradiction. On the one hand, we see the democratic moment opening up possibilities, for the first time, of popular local economic development. On the other hand, within a decade, this optimism has been transformed into a sense of fatalism and frustration. In spite of attempts to attract investment to this region the number of secure jobs has declined. Although the ANC remains hegemonic at the national level, at local government level it faces demands to meet basic needs that it has not fulfilled. The failure of the party of national liberation to respond to this challenge has resulted in the emergence of a range of semi-formal community based organizations that are attempting to fill the gap. But these attempts are merely coping mechanisms and cannot address the very real and deep desperation emerging from socio-economic underdevelopment.

Indeed, there is a sense of passivity and fatalism in the community and a lack of overall vision of an alternative response to the unquestionably deep social crisis that faces Ezakheni. However, in recent years, many have put their faith in Jacob Zuma, a charismatic ANC leader who skilfully draws on cultural symbols and Zulu identity to mobilize against the top-down style of the Mbeki government and its lack of service delivery. We will return to the implications of this type of mobilization in Part Three. First, we examine trade unions and the community in Changwon.

6

Escaping Social Death in Changwon

The Place ...

Not to have a job, it is said in Korea, is to experience 'social death'. A popular song tells the story of a man who loses his job. Instead of telling his family of his predicament, he leaves his home every morning pretending to go to work. His son sees him in a video-game arcade. He does not confront his father, and does not tell his mother or siblings of the shameful secret that he shares with his father. He does not have a job.

The city of Changwon's website shows three photographs. The first depicts a rural landscape with a mountain in the background. The second shows a massive construction site. In the third photograph, there is a modern city, with a central boulevard consisting of eight lanes – four lanes running in each direction. The only feature that remains constant is the mountain in the background.[1]

Changwon is a microcosm of South Korea's rapid economic and social transformation. The place was transformed from a rural peasant land into a huge industrial complex. The city itself was a creation of the military regime of Park Chung Hee, who announced a programme of *Heavy and Chemical Industrialization* in 1973. The small town of Changwon, which is close to the port of Masan, was targeted for the manufacturing of machine tools. The regime established the Changwon Machine-Building Industrial Complex and more than a hundred new factories were built in order to reduce the country's dependence on foreign machine tools (which was 86 per cent in 1970). The strategy was a huge success. By 1977, Korean manufactured parts accounted for 90 per cent of locally manufactured cars. Machine building in Changwon increased by 36 per cent per annum during the Third Five-Year Plan (Cumings 1997: 324–325).

The city's website states that Changwon was Korea's 'first artificial city to be developed by city planning'.[2] The city planners were inspired by Canberra in Australia, with ample parks and clearly demarcated residential and industrial areas. It was officially declared a 'city' in 1980, and it had integrated the neighbouring Dong-myeon, Buk-myeon and Daesan into its borders by 1995. The city is located at the far southeastern end of the Korean Peninsula. This is the part of the country where its military rulers came from. It is a segment of a massive industrial belt that was created by Park's regime. Popular wisdom has it that Changwon was also designed to serve as a fall-back capital should Seoul fall in the war with North Korea. In the eyes of the military rulers at the time, the Korean War which started in 1950 was far from settled when the ceasefire was declared in 1952. The city is strategically surrounded by mountains – Mount Cheonju to the northwest (656m high), Mount Bongnim to the east (567m), Mount Bulmo to the southeast (802m), Mount Jangbok (566m) to the south, and Mount Palyong (528m), also to the south. Since the city is surrounded by mountains, it can be defended from ground forces. The wide boulevards are said to double-up as runways for aeroplanes to supply the city with food and munitions in case of a siege.

In December 2005, Changwon had a population of 504,520 (257,961 men and 246,559 women) living in 168,141 households. In 1980 the city had a population of just over 110,000.[3] This means that the city's population increased nearly five times over in 25 years. Most residents of Changwon live in huge apartment blocks typical of modern-day Korea. An American who taught English as a foreign language in Korea reported on his website about his stay in Changwon. In a jab at former Cold War enemies, he described these apartment blocks as follows:

> My first sight of Changwon City wasn't a pleasant one. Some people here live in houses, but it's far more common to have an apartment ... The first-time visitor to this country could easily be forgiven for thinking that they've arrived in Communist North Korea if they were to judge by the apartment blocks. I'd say 'functional' is the best adjective I could apply to them. They're towering, 20 or 30-storey buildings, uniform, grey, rectangular, bland, and well, just plain ugly. It's like they've taken the concept of architecture and made it a pure science, not art ... The result is something that I imagine you see in the more run-down areas of Moscow, or now demolished buildings anywhere behind the Iron Curtain.[4]

The construction of the Changwon Industrial Complex was announced on 1 April 1974. It was completed in 1978. The LG factory was opened in 1980. By 1981 there were 76 tenant firms in the industrial complex. By the end of 2003 these had increased to 1,137 firms located within the complex and an additional 207 firms outside of the complex. These firms are the employers of almost 76,000 workers. The complex accounts for 3.8 per cent of South Korea's exports. Changwon is pretty much an urban industrial centre. However, some farming activities take place within the administrative boundaries of the city. There are 6,100 households, consisting of 19,300 individuals who are engaged in agricultural activities, accounting for 3.7 per cent of the city's population. The largest part of agricultural land is covered by rice fields, followed by general agriculture and some orchards. The average size of agricultural land per household is 1.7ha.[5]

Employment Relationships and the Growing Sense of Insecurity

In Changwon irregular workers (those on short-term contracts or working in outsourced companies) felt the pressure of international competition more than regular workers. For example, one irregular worker observed, 'My employment is not secure; because the LG Company is limited under global competition and my company [an internal outsourced company providing goods and services to LG only] is heavily dependent on the mother company in operation and production. The existence of the company itself is very uncertain and vulnerable to the LG's situation and external environments.'[6] Another remarked: 'The orders for production is very precarious and fluctuates. Because I belong to the outsourced company as a dispatched worker, I am so worried about my unstable employment situation. I do not know how long I can work.'[7] Or: 'LG company wants to utilize the outsourcing company with big flexibility ... If LG stops making orders from [the company] where I work, my employment agency can dismiss me.'[8] Similarly: 'I feel my employment has been very insecure. I have been an irregular worker with a monthly contract. I have had to sign my contract every month with the same employer around four years.'[9] The seven respondents who felt relatively secure were in regular employment but they nevertheless expressed uncertainty about the future: 'I feel some fears or uncertainties about my employment security, because the mother company, LG Electronics, has moved its components to China.'[10]

In Changwon, respondents identified two changes in the employment relationship that led to greater employment insecurity. First, the number of regular workers had been reduced and the number of irregular workers had increased.[11] This increased sense of insecurity was clearly reflected among respondents who had been 'internally outsourced':

> I have moved around some companies related to LG. During the past five years, I have been working in LG, but I have not been an LG worker. The regular workers of LG can enjoy better employment security than so-called 'irregular workers' like me who are working in internal or external outsourced companies related to LG. I feel my employment security has deteriorated.[12]

Similarly a regular worker in one of LG's suppliers observed: 'In our company the number of irregular workers has dramatically increased. Many jobs of regular workers have been replaced with irregular ones … They make me to be worried about my employment security.'[13] This sense of insecurity was echoed by an irregular worker in the supplier company:

> When I was a regular worker in the LG factory, I enjoyed relatively better working conditions and wages. But, now I am an irregular worker in a labour agency (outsourced company). This means I am under very vulnerable and precarious conditions.[14]

A second change that has created greater insecurity has been the outsourcing of production to China. This has affected regular workers in LG.

> For my colleagues at the workplace, employment security has deteriorated. Many workers have had to leave the company because their assembly lines moved to China or have been outsourced internally and externally.[15]

The workers in Changwon provide an interesting contrast to Orange, as many of them emphasize the importance of collective action through the trade union movement. As one irregular worker in a component manufacturer remarked:

> I think trade unions can do something to change irregular workers into regular workers. My trade union has demanded that management stop recruiting irregular workers through the labour agency company. We need to organize a kind of struggle led by the union.[16]

While this is an important observation, as it is an attempt to socially regulate the labour market, the difficulty of organizing irregular workers was stressed by one worker: 'Unity of irregular workers is urgently needed. Unfortunately, it is very difficult for irregular workers to organize themselves.'[17] Another worker remarked, 'I depend on the trade union. The trade union has strongly demanded that management guarantee employment security. Active participation in union activity is necessary. Also, work hard.'[18]

This response to working harder is of interest for two different reasons: first, the demand for guaranteed employment security harks back to the practice of lifelong employment that was a feature of the Korean labour market until recently; second, this demand for guaranteed employment exists alongside the emphasis on the need to work hard.

The emphasis on hard work is echoed in the response of other workers to the question of what measures they would take to improve employment security. A number of respondents mentioned increasing productivity, as well as encouraging their wives to earn money. These market adaptations are also reflected in the response of two other workers who wanted to start their own businesses. One said, 'I am planning to have my own shop. This means I want to be self-employed and have my own pub.'[19]

Households

While also resembling nuclear families, households in Changwon tended to consist of two parents and a child. The average size of the households of workers we interviewed was three. Households in Changwon are somewhere in between Orange and Ezakheni in their size, structure and function. Many Changwon households send money to their parents and parents-in-law or siblings. Because Korea does not have a developed state social security system, workers tend to invest in private health and accident insurance with a premium payment of 100,000–200,000 won per month (US$100–200). Although they earn good wages, the majority said that their income is not enough to meet basic needs and they have to rely on credit. Some workers pointed out that members of their households brought in additional income. 'My income is not enough', said one. 'My wife manages a small shop for supporting family livelihood.'[20] Another said that his wife earned money to supplement their livelihood.[21] Others were attempting to cut their household expenses. 'I have been economical not to use my car … Also, I have reduced the cost of basic needs like food and clothes.'[22] Said another: 'I just manage to maintain my basic life … I should be economical of my spending even for basic goods.'[23] One of the interviewees had some support from his parents: 'I am provided with basic food like rice and vegetables from my parents who are farmers in a rural area.'[24] Others are concerned about rising debt: 'My income is not enough. I have borrowed [revolving] money from credit cards.'[25]

From our interviews we generally gained the impression of workers who felt that they were barely managing. As one worker said, 'With my wage, I just manage to maintain my livelihood.'[26] The opportunities for generating income from non-wage activities in Changwon are limited. Workers tend not to own their own homes, but rent very small apartments. Only one worker rented out a room, securing a 100,000 won a month extra. Indeed, when respondents were asked what they most wanted they often mentioned a larger apartment. For example: 'I want to move to a bigger house';[27] 'I want to buy a bigger house and a plot of land';[28] 'I want to buy a bigger apartment';[29] 'I want to move to a bigger house';[30] 'I require a bigger house and a car';[31] 'I want a car and a house';[32] 'I want to move to a bigger room with more space';[33] 'I want to move from rented apartment into my own house or apartment.'[34]

Opportunities for subsistence activities are also limited, although a few do have access to land. While a number of respondents mentioned that their wives were engaged in informal market activities such as running a small shop, the dominant response to emerge from our interviews is that workers respond to the feeling of insecurity by working harder.

We were surprised by the length of time our Korean respondents work every day, week, month and year. The norm seems to be ten hours a day, six days a week. Since Korea only introduced a five-day work week quite recently, workers tend to work on Saturdays to boost their overtime income. Indeed, our respondents typically worked 11 hours overtime a week. Some in our sample, especially those who were irregular workers in component suppliers, even worked on Sundays as a rule – this is in addition to work on Saturdays! LG's own irregular workers also tended to work longer hours than regular workers in LG and the component supplier where we conducted our interviews.

If this indeed is the case, then workers in Changwon are working nearly 20 hours longer every week than those in Orange. They are also earning more than their counterparts in Orange, mainly because they boost their income by working overtime. More importantly, they are working longer hours because they are doing on average 11 hours of overtime per week. Workers tend not to take holidays and reported that they worked 12 months of the year. The concept of leisure is still something unfamiliar to the working people of Korea.

Restructuring, Political Parties and Community

The ruling political party in Korea is the centre-right Grand National Party. However, only three of our respondents belonged to a political party, namely the Democratic Labour Party. It is a small party formed to represent workers, with only nine representatives in the 299-strong National Assembly. Their support for the DLP is largely because Changwon is an industrial city (in the words of one of our interviewees, 'a workers' city') with a high proportion of workers employed in Samsung, GM-Daewoo and LG Electronics. The Changwon industrial complex was established, as we showed earlier, by the military government in the 1970s 'as a planned industrial city and as a result the infrastructure and transport is better than other cities and regions'. Workers' identification with the DLP is also linked to the fact that the KCTU launched a network for irregular workers, the Korean Contingent Workers' Centre in 2000, in association with the DLP.[35]

Unlike Ezakheni, where one party has the support of most workers, the majority of the respondents do not belong to any political party. 'I am not interested in politics and political parties' was a typical response from many of the workers we interviewed.[36] Another said, 'I am not interested in politics and sick and tired of the existing political parties.'[37] Two interviewees, both working for a component supplier, said they were too young for politics: 'Politics is too complicated to me as a young man. I do not know what politics is.'[38] 'I am too young to be interested in any political party and politics itself.'[39] An irregular worker said the following: 'No time, no interest.'[40]

Most of our respondents are not actively involved in community activities because they work too hard. A third of them listed sport – mountain climbing, specifically – followed closely by religious activities such as attending the Buddhist temple or church, and one mentioned involvement in charity activities.

As one worker remarked when asked what he would like to do: 'The community issue is too big for me to give an answer. I want to say about my personal wish. My life is so monotonous, too simple and very boring. It is a very routine life. I want to change my personal situation. I want to join a sports club like mountain climbing. I want to go up the mountain regularly for my refreshment.'[41] The two Filipino migrant workers among our respondents attend a church for migrant workers and participate in its Korean Alphabet Education Association.[42]

Most of the respondents had some knowledge of trade union activities in the community and a number of issues were identified. LG's permanent workers were aware of their union (this is the LG company union affiliated to the FKTU) collecting money for donations, as well as supporting activities for the disabled and the poor, and supporting cultural and sport events.[43] Some of the irregular workers at LG knew of some social meetings and cultural events.[44] Members of the union in one of LG's suppliers seemed to be more active, listing a range of activities. These included 'a marathon event for remembering our national division issue' organized by the KCTU regional office, as well as 'a cultural event of a song contest, [and] rallies and demonstrations focusing on labour and social issues'.[45] Another mentioned: 'The KCTU regional office organizes some meetings where I can meet labour activists to discuss pending issues', as well as 'some demonstrations and rallies'.[46] Also mentioned was the fact that 'the KCTU regional office has organized the various activities of the reunification movement, as well as political activities and sports events.'[47] Of the other irregular workers interviewed, one mentioned her union, the National Women Workers' Union, which was 'a national-level union, not affiliated to the national centres (KCTU and FKTU)'. She said, 'I belong to the Changwon local branch of the union. My union organizes demonstrations, rallies and cultural events.'[48] Another was more sceptical of union activities, mentioning a company union that was 'not active and strong'. He elaborated: 'Many union members are complaining about the current union leadership.'[49]

However, when we asked them whether they participated in trade union activities in the community only a few answered in the affirmative. It is clear that for most of our respondents, especially irregular workers, the trade union is seen as distant from their concerns in the community. An LG regular worker even felt that unions were 'too much politicized for an ordinary citizen like me to join'.[50] LG's irregular workers pointed out that they were 'not invited to these events' because they were not union members.[51] The irregular workers of the component supplier to LG seemed particularly disillusioned with trade unions. 'I don't have any network to have information on trade union activities. Actually, the existing trade unions are not interested in me as an irregular worker', said one. 'Frankly speaking, I am interested in union activity', he elaborated, 'but my employment condition is not suitable for it, because I am an irregular worker. My company does not have a union. The [company] union does not allow me to join it, because it is an enterprise union.'[52] Another did not even want to join, saying: 'I have no interest in union activity. It is none of my business. I am busy in maintaining my livelihood and household.'[53] 'I have no interest in union activity. Union is something very distant from me', said another irregular worker in this component supplier.[54] The member of the National Women Workers' Union mentioned earlier had some criticism of her union: 'Actually, my position is not consistent with that of my union. I want my union to focus on organizing in the manufacturing sector. But my union branch has focused on social and cultural events, and on organizing in the public sector.'[55] The two Filipino migrant workers we interviewed said they were interested in unions, 'but foreigners are not allowed to participate'.[56]

A key issue that emerges from these interviews is that the union movement is failing to represent the growing number of irregular workers and that these workers feel increasingly alienated from society as well as the trade union movement.[57] However, as Shin Seung Chel, the vice-president of the KCTU, pointed out in an interview: 'Organized labour is the only social actor that has an interest in protecting irregular workers. We have prioritized irregular workers because they face more exclusion and alienation from society.'[58] This is why the Korea Contingent Worker Centre (KCWC) was established with the central aim of putting the rapidly emerging 'irregular worker' issue 'on the social agenda'.[59] In his keynote speech to the Seventh Congress of SIGTUR in Bangkok in June 2005, Shin suggested that to 'counter the ideological attacks on labour it is important to change the issue of casual labour into a social issue – one that refers to discrimination including issues of gender, race and citizenship' (Shin quoted in Bezuidenhout 2005). Indeed, the issue of 'irregular' work has become a major political issue in Korea and labour law reforms that attempt to further relax regulations were stopped in parliament – for now. At the global level, struggles to fight casualization are already gaining pace, Shin argued. The ILO's discussions on the scope of the employment relationship show that there is serious engagement with the issue. Other international organizations such as the International Trade Union Confederation (ITUC), formerly the ICFTU, are also putting the issue at the forefront. In order to win these campaigns, Shin pointed out, the issue has to be presented as not only being about labour law, but about the suppression of rights. 'The casualization of labour needed to be made a global issue' (Shin quoted in Bezuidenhout 2005).

It is clear that how unions respond to the challenge of casualization is emerging as fundamental to the future of labour. 'How to cope with irregular workers', the officials from the KCWC observed, 'will be the decisive issue to decide the future of the labour movement.'[60] A two-pronged strategy is emerging in Korea to deal with this challenge: on the one hand, the KCTU is trying to reduce the number of irregular workers by fighting for the regularization of 'irregular' workers, and on the other hand, they have begun to organize irregular workers directly.[61]

This two-pronged strategy has encountered a number of obstacles. First, the KCTU has tried to persuade trade unions to open their doors to irregular workers by amending their constitutions. They have tried to persuade them to revise their collective bargaining agreements by extending them to irregular workers. But the government is eager to make the labour market more flexible by expanding the number of irregular workers and, of course, there is growing pressure from capital to make the labour market more flexible. Furthermore, irregular workers are usually excluded from labour rights. It is necessary therefore to campaign to change the labour legislation, and the institutions designed to protect workers, in ways that will make the workplace less insecure.[62]

But to organize workers, financial and human resources are required, and the KCTU has embarked on a campaign to 'collect money from regular workers to support irregular workers'. However, as Shin observes, 'Regular workers feel a natural sense of superiority. This means that you have to change their consciousness and make them recognize that irregular workers are also workers ... to harmonize regular workers with irregular workers you have to narrow the gap in lifestyle and the culture of the existing unions.' In addition to questions of identity, and especially since the 1997 economic crisis, 'there is a strong sense of insecurity among regular workers. They are now faced with unemployment and fear for their future. Who will be next, they ask.' Indeed, Shin continues, 'sometimes unions themselves make deals by agreeing to the employment of irregular workers. These deals are minority cases, but they happen. So we want to change the trend so it does not increase.'[63]

A further division exists in the Korean trade union movement, as 60 per cent of the unions are still enterprise unions. The KCTU has embarked on a campaign to build industrial unions as 'a tool to decrease tension between regular and irregular workers'.[64] The aim is to merge unions into seven 'super-unions' by 2007. However, irregular workers tend to be scattered, so it is difficult to organize industrial unions of irregular workers. The result is some tension between the organizing activities of the general unions based on regions and the existing trade unions, including those who have recently merged. They are sceptical of the idea of industrial unionism. What are needed, they suggest, are new forms of organization, not collective bargaining. 'General unions that are rooted in communities, what you call social movement unionism', they remarked.[65]

What possibilities are there of embedding unions in the community? It was estimated by our respondents that approximately forty community based organizations/social movements existed in Changwon. Very few of our respondents indicated knowledge of these community based organizations or social movements or said that they participated in them. One respondent mentioned the local Women Association and the local Youth Association, both local government organizations consisting of civilians. Others mentioned the Village of National Reunification, a nationalist movement for reunification of North and South Korea, the Association of Women and the Association of Women Workers as well as the House of Workers, a labour support organization. A further organization involved in unification is the People's Solidarity-Reunification Solidarity, a nationwide organization with local branches in Changwon.

Those who participated in community activities generally also donated a part of their wages to these causes. 'I donate money to social movement organizations. And I actively participate in the Association of Women Workers and the House of Workers',[66] said one worker. Another said, 'I donate money to this organization. Also, sometimes I participate in its events.'[67] And, finally, a worker supported an organization that strives for the reunification of North and South Korea: 'I have participated in the Village of National Reunification. I have donated 10 per cent of my wage to this organization.'[68] These were all workers in the component supplier who were members of the union that associated itself with the KCTU.

The nature of the problems identified in the community by our respondents related to the social impact of rapid urbanization in the city and the spread of irregular employment. The contrast with Ezakheni is striking. None of our respondents mentioned unemployment, as the unemployment rate in 2005 was 3.7 per cent in Korea (World Fact Book 2004: 7). No one mentioned HIV/AIDS, as the prevalence rate is less than 0.1 per cent. Deaths from HIV/AIDS are less than 200, and only 8,300 people in the entire country in 2003 were living with the disease (World Fact Book 2004: 4). Only one person mentioned crime (the growth of teenage crime) as an issue, as Korea is designated 24th out of 60 in a global ranking of per capita crime (United Nations Office on Drugs and Crime, Centre for International Crime Prevention 2005).

The responses to the question of what problems face the community were distributed as follows:

1. Housing: lack of accommodation and housing costs
2. Environmental problems: lack of public parks, uncollected garbage and air pollution
3. Employment insecurity: relocation of factories to China and irregular work
4. Transport: too many motor cars and inadequate public transport

The cost of housing is placed at the top of the list as a number of companies, such as LG, have moved their assembly lines to such countries as China, India, Vietnam, Indonesia, Russia, Thailand and Mexico and have sold their land, leading to a surge in real estate prices.[69] The high cost of land has also become a significant financial burden to the existing companies and potential investors.[70] When respondents were asked who should address these problems, the largest number mentioned local government, followed by the community and social movements. Very few mentioned the police or trade unions.

Our interviews revealed a minority of respondents who are active in the militant trade union federation, KCTU. These workers were also active in local politics through feminist movements and the struggle for national unity between North and South Korea. Most of these activists worked in a company that supplies components for LG and other larger firms, and their union activities were conducted in a quasi-underground way. They are not formally affiliated to any of the national centres, since they fear that their firm will then lose their contract from LG, which is seen not to allow unions in any of its suppliers. However, they participate in KCTU activities, and have a good relationship with this federation's local office. As the president of the local branch remarked, 'I am a movement activist as well as a trade unionist. I want to do my best as an activist.'[71]

This militancy was shared by another committed activist: 'I have a dream to live for workers. I want myself to be more involved in the labour movement, especially organizing irregular workers and raising their consciousness.'[72] These views come closest to what we have called a social movement approach, in that they see the trade unions working together with the community based organizations to solve the problems facing society. The perspectives of this grouping are best captured in these three responses:

The trade union should play the most important role to resolve these problems with the cooperation of civic groups. The wage gap cannot be solved only by government measures. The rich people should abandon their wealth to the public. Who can enforce the rich to do it? I think the role of union and social movements is very significant.[73]

I believe that united struggle and solidarity consciousness can pave the way for resolving community problems.[74]

In order to resolve the issue of irregular workers, I think the role of national and local government is the most decisive. The governments need to improve political and legal institutions for irregular workers.[75]

Conclusion

As a working-class city, Changwon has experienced the threat of down-scaling and relocation of sections of industrial production to China. The national project (the unification of North and South Korea) remains incomplete, as North and South remain divided. Workers on the left of the labour movement feel strongly about the issue, and participate in reunification organizations. They have also put their faith in the KCTU and the Democratic Labour Party.

The Korean case stands out as the clearest example of a trade union movement engaging innovatively with the question of irregular work and attempting to socially regulate the labour market. While casualization is clearly a global phenomenon, it has a special salience in Korea as this is where, until recently, the large companies such as LG practised lifelong employment among their core workers. The rapid growth of irregular employment, as well as the relocation of parts of the production process to China, is a dramatic threat to the rights of core workers established in Korea. This is generating an intense and militant struggle between labour and capital, with the state under increasing pressure to create a more flexible labour market. Whether they succeed is unclear but, as emerged in the seventh Congress of SIGTUR in June 2005 and then in the strike in November 2006, the Korean labour movement is determined to define labour law 'reform' as a 'social issue' that undermines hard-won labour rights laid down by the ILO Conventions. Importantly, it is also a struggle that they intend to globalize. Apart from these struggles directly related to the world of work, the issue of national unification is a major impetus to militancy among Korea's labour left in Changwon.

Nevertheless, the dominant response among the workers we interviewed to insecurity was to work harder and to increase their overtime work. Furthermore, in the absence of a universal state social security system, workers are responding to household insecurity by investing in private health and accident insurance schemes. This is in sharp contrast to Orange, where workers are being squeezed by the forces of corporate power.

7

Squeezing Orange

The Place . . .

In their song 'Truganini' the Australian band Midnight Oil sing about their country as a road train on its way to nowhere, where the 'roads are cut' and the 'lines are down'. They sing about environmental degradation, farmers who 'are hanging on by their fingertips' and about blue collar workers who are caught in a debt trap. In the lyric, they address working people directly: 'Somebody's got you on that treadmill, mate... And I hope you're not beaten yet.'[1]

In this chapter we examine the impact of the restructuring of the Electrolux factory on working people in Orange in New South Wales. In 1999 Defy's current operations manager in Ezakheni visited the Orange plant. At the time he was working for Kelvinator in Johannesburg and the factory in Orange was owned by Email, the local Australian corporation. Email was considering taking over Kelvinator in South Africa. He told us that he was impressed by the Email operation – he considered it to be a highly productive plant. He sensed Australians liked their sport and that the plant was 'heavily unionized'. Moreover, Email was able to compete against its competitors, and they were proud of this. Indeed, when we mentioned that Electrolux had bought the plant in Orange and that they were in the process of downsizing the operation, he seemed genuinely surprised.

When Electrolux took over the plant in 2002, they brought a group of Korean workers from the LG factory to visit Orange. Leon Adrewartha, the director of manufacturing for Electrolux Australia, explained how the company viewed the Korean work ethic:

> I mean those people [Koreans at LG], when I say work, they work. They're doing exercises in the morning not to feel good, it's so they can work flat out for eight to ten hours a day for six days a week. How do I create an environment that lets us work at the same rate and tempo that they do?[2]

Orange is a company town. White goods manufacturing is central to its local postwar history. We began our research here in 2002, following the Electrolux buyout. To be sure, the production facility in Orange was one of the most significant manufacturing sites in the country. Also, the factory is central to Orange's local economy. When the Census was conducted in 2001, there were 2,186 local manufacturing jobs. This accounted for 14 per cent of the 15,425 jobs reported to the statistical authority at the time. Roughly 1,800 of these manufacturing jobs were in the Electrolux plant. As we showed in chapters 2 and 3, the company has downgraded the manufacturing facility to an assembly operation and slashed the number of production workers to a mere 450 individuals, of whom 100 are employed as casual workers. The local impact has been devastating. As a worker in this factory said in an interview:

> We are now just a clock number; we're just a tally to get out. That is what Electrolux is about. That is how we feel.[3]

To get to Orange, you have to travel for 260 kilometres from Sydney in a westerly direction on the Great Western Highway, across the Blue Mountains. Orange City is characterized by its rural nature. There are expanding residential areas, as well as some land used for industrial and commercial purposes – notably the Electrolux factory. Its administrative boundaries encompass a land area of about 290 square kilometres. Of this, 90 per cent is rural land, mostly used for forestry, mining, sheep and cattle grazing, crops, orchards and viticulture. Most residents live in the city of Orange, but there are also smaller villages, including Lucknow and Spring Hill, as well as even smaller settlements such as Huntley, March, Shadforth and Spring Terrace.

The area was originally inhabited by Wiradjuri Aboriginal people. However, in 1822 Captain Percy Simpson established a convict settlement called Blackman's Swamp. John Blackman, the Chief Constable, was a guide who had accompanied Simpson and another explorer before him into the region. The local union delegates were aware of a massacre of Aboriginal people that took place where the town square is today. According to them, there were massacres in other surrounding towns as well. In 1846, Major Thomas Mitchell renamed the settlement Orange, in honour of the Dutch royal Willem van Oranje (William of Orange), whom he had met during the Napoleonic War. Like Ladysmith in South Africa, the discovery of gold also impacted on the town. In 1851 gold was discovered at nearby Ophir, and the resulting Gold Rush led to an influx of migrants into the area. Orange became a central trading area. Because of good agricultural land and weather suited for farming, Orange also became an agricultural hub and was proclaimed a municipality in 1860. A few years later, in 1877, the railway from Sydney reached the town. This contributed to the growth of Orange.

By 1933, Orange had a population of 7,700. The most significant development occurred in the postwar years, with the population doubling between 1946 and 1976, to nearly 30,000. During the 1980s, population growth slowed down, but picked up again in the 1990s. The population has steadily increased over the last ten years, from 35,000 in 1996 to its current size of just under 38,000.[4]

Today, the local tourism agency markets Orange to residents from in and around Sydney and other visitors as 'a home away from home'. 'It is easy to find your way around Orange. You'll feel like a local in no time', a promotional video at the information bureau tells us. The video promotes local history, wildlife, food and 'cool climate wines' now produced in the vicinity. It boasts a civic theatre and art galleries, as well as 'quaint shops' at the historical Millthorpe village. The town is also targeting Australia's growing cohort of pensioners. Local hospitals and healthcare facilities are being upgraded.

In short, there is an attempt to build a town that is not a company town – a post-industrial settlement, where the mining industry is part of a manufactured nostalgia, and not the environmental destruction caused by modern opencast mines. This nostalgia has little tolerance for factory workers or massacres. Old Victorian working-class homes are fixed up for urban yuppies who want to escape from the city for a weekend. McMansions, as the new double-storey houses are called in the local vernacular, are rising like mushrooms next to the botanical garden at the edge of town.

The Employment Relationship and the Growing Sense of Insecurity

In Orange all the workers we surveyed in 2005 felt insecure about their possibilities of securing long-term employment. One worker remarked,

'The company is putting people off left, right, and centre; they are closing sections down. I don't know how long I will be employed.'[5] Another observed, 'It has become worse with free trade agreements; you cannot compete with lower wages in other countries, which makes their product cheaper.'[6]

In Orange, workers displayed a range of responses with both fatalistic reactions (there is nothing you can do) and market adaptation (additional training or looking for another job) predominating. However, these workers also hinted at collective resistance to liberalization by suggesting that workers should only buy Australian products or, more significantly, make it cost-effective not to outsource.

The company's capacity to command and produce space, to use geography to reinforce its structural domination over labour, has further undermined workers' belief in the value of unionism, for unions seem paralysed by these changes. Fatalism pervades Orange. An internal memorandum from union organizers in the town argues that a majority view the loss of their jobs as inevitable. They do not believe that they have the capacity to 'turn it around' with only a 'small minority' being 'prepared to take industrial action'. The majority feel betrayed by the company. They had turned their back on the union and sided with the company during the enterprise bargaining process only to discover that Electrolux had abandoned them. Organizers believe that the majority have become 'very conservative and are unable to comprehend the implications of free trade and global competition on their day-to-day lives in Orange'.[7] The response of organizers is also fatalistic, with the primary focus on redundancies thereby signalling that the restructuring decision cannot be challenged. The organizers also noted that the restructuring will undermine the viability of the union branch because of reduced membership. They doubt the 'long-term viability of the plant' due to its reduced status in the global production chain and Electrolux's global capacity to absorb what will be left of Orange over the next three years.

Our interviews reveal that workers experience the new work regime as a psychological shock, tearing at the material basis of the lives that they had constructed in this country town. They are divided between their feeling of belonging in Orange and their experience of restructuring which creates profound insecurity, a condition which erodes a sense of well-being. For example, some of the Orange workers that we interviewed spoke enthusiastically about the small plots of land that they had acquired by taking out a mortgage. Yet in the next breath they would articulate their fear that the restructuring would end this lifestyle that they had come to enjoy.

Certainly, the geographically based, structural domination of capital, reinforced by the Australian state's anti-union laws, has undermined belief in an alternative. Workers already speak nostalgically of life in the town before the Electrolux onslaught, when they felt relatively secure in a meaningful place, believing they had lifetime employment. Now their world has changed, seemingly irretrievably. We returned to Orange in late April and early May 2005 to discover that the overwhelming ethos of workers in the factory and those who have been dismissed is a paralysing pessimism: nothing seemed to stand in the way of global corporations who can restructure at will. Speaking to workers at the factory gates during an afternoon change of shift, this viewpoint was repeated: 'What happens, happens, there's nothing I can do.' In an interview, a union activist commented, 'No one gives a fuck anymore; they just experience Electrolux as a big company that has just taken over their lives. Everyone knows that they are just a number now. That's why they don't give a shit about anything.' Then there is Sarah, a union shop steward, who had been a fighter. On a previous field trip in 2004 at a union rally, she was articulate; a fiery person, unafraid to speak her mind. At a local council meeting the night before the rally, she stood up and attacked councillors who were real estate agents, and who had argued that there should not be a fuss over the threatened closure, as this might affect their businesses. She warned, 'Just remember, the community elected you and the community will chase your fat arse at the next elections.'

A year later in May 2005, Sarah was unrecognizable. She had shifted from fighter to fatalist. She said, 'There is nothing that you can do. You just have to move on. I've now got a low-paid job as a cleaner in a hospital. I just want to move on – you can't do anything about a big company like Electrolux.' In a real sense there was nothing that she could do, because the unions had done nothing other than routine activity, caught as they were in an institutional groove that was effective in the age of national unionism and bargaining rights. There are also gender issues here, as Sarah's partner is against her becoming involved in any campaign against such an 'important company' in the town. Others we interviewed echoed Sarah. Dennis said, 'Life has just got to go on. You have got to make it go on.' He is a victim of the 'downsizing exercise'. Stressed and anxious, he reflected, 'I have got nothing in my bank account. Centre Link won't pay me now – they say I will get my first payment at the end of May. I will try and live on my A$1,500 balance on my credit card.'[8] With no prospect of collective resistance, workers have shrunk into their private world of personal struggle, a daily grind just to survive. Patric reflected, 'I have been kicked in the gut. I don't have it in me to fight anymore. I just want to take my A$50,000 redundancy payment and run.'

According to a union delegate, who was still working at the factory when we conducted our last research visit to Orange in November 2006, most of the workers who were laid off were actually 'glad to see the end of Electrolux'. Those who did not move elsewhere, found jobs. But like Sarah, most workers got jobs in the service sector, where wages are lower and conditions are less favourable. According to the union delegates, a new opencast gold mine has recently opened up in the area and could last for about twenty years. In reality, Orange has become a microcosm of the broader structural shifts brought about by trade liberalization in Australia. Manufacturing closes down, shifting workers into the service and mining sectors. Australian unionists often talk about their country's new status as 'a quarry for China', referring to the resources boom that led to the mining renaissance in the country. The downside, of course, is that a strong currency built on the back of resource exports further undermines the competitiveness of manufacturing.

Despite this bleak picture, commitment from a minority to resist coexists with this inward turning pessimism and passivity. While Electrolux has used downsizing to rid the company of union delegates, remarkably, those who have survived the purge are determined to fight. A tattooed bikey[9] delegate (or shop steward) nicknamed 'Hungry' refuses to kowtow to management and supervisors and their anti-union campaign, proudly wearing his shirt with a bold union logo. He is an archetypical Aussie worker, humorous, upfront and honest. 'I'm not afraid of them; they don't scare me; I'm with the boys.' Then there's John, who has steadfastly remained a union member for the past 20 years. 'The union saved my job on two occasions. I will never leave the union.' Brad has had enough of Electrolux's ways. He has suffered from serious repetitive strain injury (RSI) for the past three years after the speed of the assembly line accelerated dramatically. 'We are just a number now, just a clock card. There's no human side now – just push, shove, hustle, go, go, go! They have jacked up the speed by about two thirds and workers are finding it difficult. The injury rate has gone up.'

The remaining delegates form a small nucleus of committed activists who are prepared to risk everything rather than submit to the corporation. These leaders are a classic illustration of the optimism of agency, confirming Herod's (2001a: 5) view that 'there is always opposition to power and domination' in the face of 'the juggernaut of global capital'. Their brash, fearless opposition to the company's plans confirms Harvey's argument that the 'transformative and creative capacities' of persons 'can never be erased' (Harvey 2000: 117). His contention that persons are 'the bearer of ideals and aspirations concerning, for example, the dignity of labour and the desire to be treated with respect and consideration as a whole living being' certainly comes into play. For example, there is a universal hostility to the dramatic increase in line speed that is proving so damaging to their bodies.

Despite the existence of this opposition within the factory, questions remain. Given the hostile anti-union environment, how can this small, beleaguered group transform the fatalism of the majority? Considering these questions was a crucial juncture in the research process when we came to the obvious conclusion: left to their own devices and armed only with the classic union response to downsizing and future closure, namely, nothing can be done except to negotiate redundancies, their opposition to restructuring would remain largely ineffectual.

This predicament challenged the union's national leadership – would they simply concede defeat, or could they envisage a strategy that would give new direction and hope to this minority? This is a struggle which has taken place within the union leadership between opposing perspectives that believe that resisting restructuring is a lost cause and challenging those who believe an alternative is possible. We will return to this question towards the end of the chapter.

Households

The unemployment rate in Orange was 7.3 per cent when the 2001 Census was conducted; 61.7 per cent of those older than 15 years worked in full-time jobs, and a further 28.6 per cent worked in part-time jobs. We can see how the shift to insecure forms of employment has already impacted on the local community. Compared to people living in Ezakheni, Orange does come across as an affluent society: 90 per cent of households in Orange owned cars; 40 per cent owned one car, 30 per cent owned two cars, and a further 10 per cent owned three or more cars. The overwhelming majority of working people in Orange drive to work in their own cars (62.5 per cent); 83.4 per cent of residents own the houses they live in, with the rest in townhouses or apartments.

Orange is also a relatively homogeneous community. The overwhelming majority of the residents speak English at home (92.4 per cent), with speakers of minority languages including Italian, Russian, Polish, Tagalog, Croatian, Greek, German, Arabic, Spanish and Chinese. Those who do not speak English as a first language add up to just over 1,200 individuals. The overwhelming majority of Orange residents were born in Australia (87.4 per cent), with half of those born in foreign countries coming from mainly English-speaking countries.

The households of the workers we surveyed in Orange are typical nuclear families, quite often with only one or two people. Indeed, the average size of these households was 1.6. According to the 2001 Census data, the overall household size in Orange is 2.59 (see table 7.1). Many of the workers we surveyed were young couples without children, which explains the lower average household size.

In Orange, households are dependent for income largely on their wages and rely on state transfers in the form of unemployment benefits and their pensions if they are laid off or facing early retirement because of closure. Although these benefits are being eroded, Australia's welfare state provides these workers with a reasonably comfortable alternative to wage employment. Furthermore, they are not obliged to provide an income to any other persons than their immediate nuclear family. Although they earn relatively good wages (A$700 a week on average), they save very little money monthly and spend most of their discretionary

Table 7.1   Household composition in Orange, 2001

Households	Number	%
Couple without child(ren)	3,088	24.1
Couple with child(ren)	4,272	33.3
One-parent family	1,583	12.3
Lone-person households	3,252	25.4

Table 7.2   Household income in Orange, 2001

Income	Number	%
Weekly income under A$300	1,765	13.9
Weekly income A$300 to A$999	5,436	42.9
Weekly income A$1000+	4,167	32.9

income on consumer goods and sporting activities. Table 7.2 represents the average weekly income of residents of Orange. Respondents said they work a 40-hour week and occasionally do overtime in the season. This conforms very much with the average for working men in Australia, although if women and men are calculated together the average of working hours per week drops to 35.6.

In Orange we can see the results of a century of social democracy. Australian people believe that they have the right to a life of recreation outside the workplace. The question is whether this lifestyle can survive the longer-term effects of global competition facilitated by the intro-duction of extremist anti-union labour laws in December 2005, which facilitate the cutting of material conditions and the flexibilization of working hours.

Restructuring, Political Parties and Community

Orange represents a significantly different response to that of Ezakheni as no one belongs to, or participates in, any political party. The feelings expressed were ones of betrayal by the Labour Party in Orange because it had introduced market liberalization which has had such an impact on work and lifestyle. The ALP's economic policy has also produced endless restructuring that has changed their way of life. Furthermore, respond-ents evidenced a low political awareness except for some support for One Nation, a right-wing party that won some support in the 1990s for its anti-immigration policies and its desire to return to a more protectionist trade policy (Kingston 1999).

One Nation's leader, Pauline Hanson, emerged, according to veteran journalist Margot Kingston, 'as though she was the hidden underbelly of the Australian psyche, throwing up ideas and values most of us had thought buried or even gone' (Kingston 1999: 4–5). Although the Australian electorate eventually firmly rejected her and her party, she did strike a chord among ordinary workers when she spoke 'of the evil of economic rationalism' and the need to keep Telstra in public hands in order to save jobs, and criticized corporations for undermining small enterprises. In Hanson's own words:

> And I'm not for this free trade, and I've spoken about that for the last couple of years, and just recently Bill Clinton's done a turnabout on it. We hear Dr Mahatir is actually doing a turnaround because he says now that

free trade is destroying his country. Is it all right when they say it, but when Pauline Hanson says it I'm wrong or I'm simplistic? (Quoted in Kingston 1999: 63)

Underlying her popularity among what Australians refer to as the 'battlers' was a xenophobic racism, as illustrated in this quote by one of her classically insecure and frustrated young Labour blue collar voter turned One Nation supporter:

> She talks what she believes in and not what everyone else wants to hear ... And the way I look at it now, being eighteen, is she's saying basically along the same lines: the Asians are coming over here and taking the Australian jobs from the Australian youth. And there is just nothing left for us to take. It's like we have to fight doubly harder, and if we take that job we've got to take the pay cut rather than taking the proper pay that we should get, because you've got the Asians coming over here and they work for next to nothing, basically. Which is really hard for a bloke like me. My dad, he was on the hierarchy on the State Rail Authority and so was my grandfather before that, and my uncle. But for me, the railway is a dying breed, it's being run from computers. (Quoted in Kingston 1999: 175)

Ironically, Orange used to be a Labour stronghold until independent Peter Andren was elected, but those we interviewed indicated that political parties had no active support. They did however suggest that what was required was more state intervention to protect local industry. One stressed the need for an alternative approach to politics and supported the independent candidate, Peter Andren.

What was striking about our respondents in Orange is their active involvement in a wide variety of outdoor sports; in 'footy' clubs (Australian slang for football), in cricket clubs, in golf, in touch football, in horse riding and hunting (the Ulysses Club), and in bowls. What is of especial interest in this comparison with Korea and South Africa is how Australian social democracy emerged in a way that made sport accessible to all, including working people. Australian identity became associated with the idea of 'this sporting nation' and some of the greatest sporting heroes came from working-class families. The key concept in developing the notion of a sporting nation was the traditional slogan of the 8-hour working day. Indeed, Australia was the first country to win the campaign for the 8-hour day in 1881. The campaign was built around the slogan of 8/8/8 – 8 hours for labour, 8 hours for sleep and 8 hours for leisure.

Table 7.3 Comparing violent crime in Orange with the national average 2005

	Times the national average
Forcible rape	2.08
Robbery	1.11
Aggravated assault	1.93
All violent crime	1.71

Table 7.4 Comparing property crime in Orange with the national average 2005

	Times the national average
Burglary	1.66
Larceny or theft	1.48
Arson	1.28
All property crime	1.42

Source: Cityrating.com

However, this cosy, secure life centred on the Rotary Club and the exclusively male Masonry is under threat in Orange. Although it is a small town, both violent crime and property crime are above the national average (tables 7.3 and 7.4).

Our respondents saw the trade union as involved in the community, organizing rallies from time to time, having meetings, running courses and providing Christmas dinners for the local pensioners. But we have identified a sense of fatalism over the future of Electrolux in Orange leading to a struggle inside the AMWU's leadership. On the one hand, there were those who said there was no alternative to the gradual downsizing of the plant and its relocation to China and the union should concede defeat. On the other hand, there was a minority group of activists willing to resist and who were looking to the AMWU leadership for a new approach to restructuring in the era of neoliberal globalization. The intra-organizational struggles inside the union are illustrated in figure 7.1.

This approach to restructuring involves the idea of a global movement that will network workers in Orange with workers in Electrolux worldwide. Figure 7.1 illustrates this debate within the union leadership nationally and workers on the shopfloor for and against a fatalistic

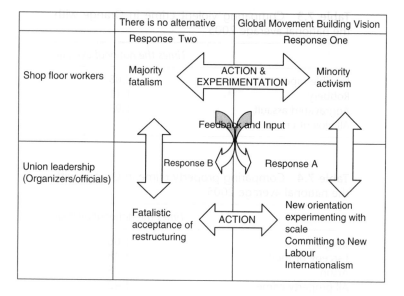

Figure 7.1 Cycles of transformation: intra-organizational struggles for hegemony

acceptance of restructuring. There are middle-level leaders (local union organizers) who argue for acceptance (response B) by highlighting the fatalism of the majority of workers in the factory (response two) to confirm that there is indeed no alternative. They contend that any attempt to resist would simply be a waste of scarce union resources. Eventually, response A, promoted by senior national AMWU leaders, won the day and a new strategy was formulated, which we briefly outline below.

The new initiative is grounded in the activism of the minority (response one) and gives encouragement and ideas to their action. If response A had failed to assert its hegemony within the union, this response would have dissipated, with others coming to share the disillusionment of past activists such as Sarah.

Spaces of Hope

Linking the local to the global in Orange involved a number of experiments, some of which failed. During the early phases of restructuring, the AMWU, through the International Metalworkers' Federation

(IMF), contacted their Swedish counterparts, who then sent three delegates to visit Orange. These Swedish trade unionists, however, deeply disappointed their Australian colleagues. It turned out that they were worker representatives on the company's board and were highly integrated into management's philosophy. They supported the restructuring and resisted any attempt to form networks that could challenge corporate unilateralism. This experience also showed the limits to World Company Councils in their existing format.

The union then changed its tactic and experimented with a more innovative approach. Response A crystallized after months of debate among national leadership of the AMWU in April 2005. A model of global unionism emerged that has been formally adopted as policy. The form of this global unionism is captured in figure 7.2.

'What happens', David Harvey asks, 'when factories disappear or become so mobile as to make permanent organizing difficult if not impossible? ... Alternative modes of organizing must then be constructed' (Harvey 2000: 56). The AMWU leadership's response to the crisis in Orange answers this question. Transforming union scale through networking is the essential dynamic basis of global unionism. In the Orange experiment, networking is by definition non-bureaucratic

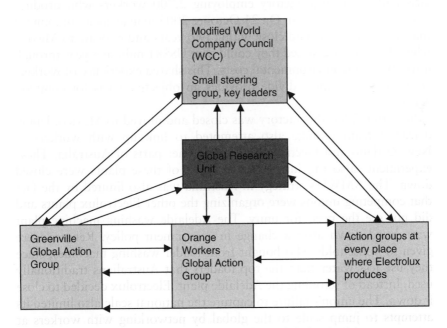

Figure 7.2 Proposed organizational model for global unionism

and has the potential to revitalize unionism, for the network presents an opportunity for direct grassroots involvement. Factory workers do not have to have the permission of a union organizer to take the initiative: they simply act, then report and seek to win union commitment to new proposals. Networking creates 'spaces of hope' by linking workers across the globe in the same corporation, giving them the opportunity to reflect together and search for ways of applying pressure to block predatory corporate restructuring. The network can build dynamically outwards from the committed minority that exists in the workplace in Orange. These leaders then become the local action group, the node in the network, connecting to other action groups across the global corporation.

The constitution of such an action group in Orange and their empowerment through developing internet skills and creating a website constructively channels their anger into a movement-building project. Initially, they moved in two directions. Their first step was to establish internet contact with Electrolux workers in the United States. The similarities between the factories in Orange and Greenville, a country town in Michigan, are striking. Both are Second World War converted munitions factories; the viability of both towns is tied to the plant; both have been bought out by Electrolux and then threatened with closure. Greenville is a large factory employing 2,700 workers who produce 1.3 million fridges a year. On 21 October 2003 management announced that the company would close within two years and relocate to Mexico where Electrolux claimed they could save US$81 million a year through lower wage and environmental costs. This shared experience of workers across the geographic divide provides an objective basis for common action.

In March 2006 this factory was closed and moved to Mexico. Union delegates from Orange also attempted to link up with workers in New Zealand and Electrolux plants in other parts of Australia. These experiments also ran into crisis when all of these plants were closed down. The AMWU's strategy in Australia was also limited by the fact that competing unions were organizing the other Electrolux plants and did not see the need for unity. The Adelaide washing machine plant was closed down after a change in government policy. Rebates were given to households who bought front loader washing machines, since they use less water than the top loaders that Australians traditionally used. Instead of retooling the Adelaide plant, Electrolux decided to close it down. The union's failure to capture the national scale also limited its attempts to jump scale to the global by networking with workers at other Electrolux plants.

A second strategy proved to be more successful in terms of immediate results. This strategy involved the creation of a coalition in Orange through involving local groups in the campaign against Electrolux. Immediately after Electrolux took over the factory from Email, workers organized a protest rally in town. Building on this experience, a Community Forum was set up and another mass meeting was held in June 2004. This was supported by local church groups, and a local Catholic priest spoke about the devastating effects of restructuring on households and living standards. Peter Darley, leader of the local apple farmers, supported the campaign when he spoke at the town rally. He argued, 'We know the Electrolux workers. Your children pick apples for us. We are family farmers. Because of free trade, our jobs are as much on the line as yours. That's why we must unite.' Apart from the farmers' interest in maintaining a strong local economy, they empathize with the workers' plight because of their own struggles against free trade and the threat of importing fire blight disease (a bacterial disease, particularly destructive of apples) into the local orchards. Their campaign slogan is 'Fair trade or failure'. Furthermore, they are engaged in battles with Australia's two large supermarket chains, who they argue are exploiting farmers through their duopoly control on pricing.[10]

Reflecting on these initiatives in November 2006, union delegates felt that the Community Forum had a real impact. The company knew that its actions were being scrutinized publicly, and this made it more circumspect in its approach. They felt that the company was also being somewhat cynical when it outsourced some of its supply functions to local operations, including a protected labour agency. Later, when attention was elsewhere, it applied the 'China price' and shifted its contracts overseas. We asked an official at the local tourism bureau about the status of the factory. She said that the plant was still operational, but that it had been downgraded to an assembly operation. We remarked that she was well-informed, and a local union organizer responded: 'Of course she is. We were bloody well in the news for six months!' To be sure, the result of the union's campaign and the Community Forum was that the restructuring became a subject of public debate. This limited the company's unilateralist approach to some degree.

These initiatives create 'spaces of hope', giving direction and purpose to resisting the negative consequences of restructuring. Agency is created wherein the inner turmoil of victims is transformed through movement building. Our research captures this ongoing process through interviews and observations. Even in this early stage, commitment to place and opportunities for *local* action are indeed created through global networking with workers in other places where they are similarly trapped and in need of an alternative strategy. Contrary to the conclusion drawn by Gibson-Graham, this is not merely 'a left vision of power' which neglects 'opening the local as a place of political creativity and innovation' (Gibson-Graham 2002: 50, 53). Through drawing the small beleaguered group in the factory into global networking, they discover new possibilities of resisting restructuring. Political creativity is already evident in the new global union model developed by the AMWU, a model which has the potential to challenge the dominant discourse and explore areas where the power of global corporations can be tested.

The creation of a global research unit could play a significant role in empowering the network through monitoring the corporation's global strategy and through challenging its restructuring discourse. Corporate discourse obscures the real nature of restructuring, hence closures are 'integrated production systems'; downsizing, casualization and outsourcing are 'manufacturing modernization'; the shift to authoritarian supervision through not allowing workers to speak on the line is 'team building' and work intensification is 'achieving targets'.[11]

Conclusion

Of the three towns we studied, Orange stands out as a company town that was built on a successful white goods industry that is now being undermined by the destructive impact of globalization. For most of the workers in the plant, there is no alternative and behind the facade of quaint tourist shops and al fresco dining lurks a sense of despair. But we have also seen in Orange the emergence of an innovative attempt to protect society against the unbridled power of the multinational corporation. The union successfully involved other local groupings to support their protest, and in doing so, raised public awareness. Furthermore, a minority of activists in the Electrolux plant have linked up to the national leadership of the AMWU to begin a process of building a global union response to Electrolux. This response is still in an experimental phase and has run up against the sheer magnitude of plant closures, but the AMWU is committed to finding innovative ways of shifting scale from the local to the global. Workers at Electrolux know that they are on a treadmill, as Midnight Oil point out, but at least some of them are not beaten yet.

Part III

Society Governing the Market?

In all three research sites we have identified two types of societal responses to rapid market liberalization: on the one hand a retreat from, or an adaptation to, the market, and on the other hand mobilization to challenge the market. We begin with retreat and adaptation.

We identified in Orange a hankering after protection against foreign imports and competition. In some interviews respondents expressed xenophobic responses to immigrants and a nostalgic desire to support right-wing parties. Most felt fatalistic about the future of the plant and intended to fall back on welfare. Changwon, a working-class city, was also experiencing the threat of downscaling and relocation of industrial production to China. Here workers responded by intensifying their work in spite of attempts to limit the length of the working week through labour law reform. Overtime increased and workers were responding to the threat of downsizing by investing in individual insurance and pension schemes. Retreat from the formal market was most striking in Ezakheni, where workers responded by pooling their limited resources in households that have grown in size while at the same time searching for non-wage sources of income.

But workers have not only responded by retreating and thus implicitly affirming the power of market relations, they are also mobilizing against markets and challenging market power. This is most striking in Orange, where attempts have been made to build global alliances with Electrolux workers to challenge the power of the multinational corporation. Of particular interest is the attempt by the union to form an alliance with local apple farmers who are also threatened by cheap imports. In fact many supported a local candidate in the recent elections instead of their longstanding support for the ALP. In Korea a Democratic

Labour Party (DLP) has been formed, and an aggressive campaign has been launched to organize irregular workers. The clearest challenge to market relations was the successful blocking of the privatization of the electricity utility.

At Defy in Ezekheni workers were able to block the introduction of labour brokers, although short-term contract work is widespread. The ANC remains hegemonic at the national level, but at local government level it has experienced demands to meet basic needs that it cannot fulfil. The result has been the emergence of a range of community based organizations (CBOs) that are attempting to fill this gap as well as a grassroots, populist challenge led by the ANC deputy president, Jacob Zuma, against the top-down modernization project led by its current president, Thabo Mbeki. Of the three countries, the most significant challenge to the global market has been the introduction of a quota on Chinese clothing imports introduced in late 2006 by the South African Department of Trade and Industry to protect the local clothing industry.

We capture these different responses in the table below.

Summary of societal responses

	Retreat/adaptation	Mobilization
Orange, Australia	Fatalism and reliance on state welfare. Support for right-wing parties. Migration to mining areas	Attempts to build global alliances. Links with farmers and the local community
Changwon, Korea	Work intensification. Individualized investment in heath and social insurance	Formation of DLP. Campaigns to organize irregular workers. Privatization of electricity utility stopped
Ezakheni, South Africa	Retreat into the household and informal non-wage income activities. Migration to urban centres	Prevention of labour broking in the factory. CBOs at the local level. Mobilization around Jacob Zuma as a populist leader. Regulating Chinese imports

In all three sites there are innovative attempts by local community organizations and the trade union movement to search for security and ways of protecting society against growing commoditization. However, these responses lack an overall vision of an alternative response to the challenge of globalization. As we argued in the introduction to Part Two, retreat is predominantly an individual response, and mobilization a collective response. The distinction between the two lies in whether they are adapting to market power or challenging it. A successful challenge to market power has to involve some notion of an alternative to current power relationships. This is the focus of Part Three.

As we suggested in our preface, the journey on which this research has taken us has led us to see the contemporary world through new eyes. In chapter 8, we argue that history matters if we are to explain the different societal responses in the three countries. We are especially concerned to explain how it came about that democratic unionism has, and continues, to play an important role in South Africa and South Korea. As we argued in chapter 1, the working class is not some *tabula rasa*; it has its own distinctive political traditions, pre-existing networks and organizational forms that shape its responses. In Australia this was labourism. In Korea there is the legacy of enterprise unionism. In South Africa it was the explosive force of national liberation.

History has also shaped the relationship between the market, the state and society in different ways in the three countries. In Korea the state dominates society with the market as its handmaiden. In Australia, in spite of its history of successful social regulation of the market, the outcome has been the opposite; the market is beginning to dominate society with the state as its handmaiden. In South Africa the outcome has been equivocal. Here neither the state, nor the market, nor society clearly dominates. There is something of a class stalemate, with a social liberal outcome – social on account of the important extension of social grants to the poor, liberal due to its adherence to a neoliberal view of how the economy should be run. South Africa holds out the promise, through the strength of its labour movement and the multipartite economic and social forum, NEDLAC, created largely by labour, to socially regulate the market.

However, as all three countries have been drawn into the global neoliberal project, the configuration of these spheres has changed, leading to ongoing contestation between the international financial institutions and global corporations as they try to assert their hegemony over the local market. The result of these global pressures has led activists to frame their action increasingly at the global level.

In chapter 9 we examine the emergence of transnational activism and show how labour has begun to work space through engaging scale from the local to the nation-state to the global level. We identify the emergence of new sources of workers' bargaining power – what we call moral or symbolic power – where unions and community activists have begun to mobilize around the discourse of global justice, fair trade, fair employment and, in the case of South Africa, access to anti-retroviral drugs in the treatment of the AIDS pandemic. Various new initiatives are currently underway in which logistical power is being carefully and strategically evaluated. The challenge, we suggest in this chapter, is for labour to go beyond the workplace and to form links with communities – what we call social movement unionism, which moves outwards into communities, while also working space through networking place to place within the same global corporation.

In chapter 10, we catch sight of our destination: the construction of a counter-movement. The problem with Polanyi, we suggested in chapter 1, was his assumption that the counter-movement was constructed spontaneously. We begin in this chapter by identifying the local experiments and institutional innovations, as well as the links between the local and the global. We then start to construct from the local a vision of what a democratic global alternative could look like in which society governs the market. It is a utopia, but it is realistic – what has been called real utopia.

8

History Matters

Australia, South Africa and Korea followed different trajectories both in terms of how markets were developed, and how society has responded to these developments. The implications are obvious: history matters. In this chapter we demonstrate how the state has intervened to structure the labour market in different ways; in South Africa, through colonial conquest and forced proletarianization, a labour market was created which has been, and still is, marked by historically entrenched racial divisions. In Korea the state disciplined capital and segmented the labour market, laying the foundations for its successful emergence as a 'late industrializer'. In Australia the state intervened at the beginning of the twentieth century to construct a class compromise through a method of judicious determination in centralized wage-fixing and the conscious protection of jobs and industry.

To understand these different developmental paths it is necessary to delve into the history of these three countries and their different political traditions. In South Africa a long tradition of political engagement emerged that was driven by labour's enduring interest in liberation and development, initially labelled political unionism and later social movement unionism. In Korea the trade union movement was tightly controlled by the state under a military dictatorship, but this form of state corporatism was challenged by the emergence of a democratic labour movement after the strikes of 1987. In Australia trade unions were a recognized, integral component of the class compromise and came to play a key role in securing a steady, uninterrupted improvement in wages and conditions, consolidating its own distinctive social democratic 'labourism'.

We have shown how, under the impact of economic liberalism, all three countries embarked on a process of restructuring introducing new modes of what Harvey (1989) has called flexible accumulation. Restructuring has impacted, in turn, on the three distinctive trade union traditions, eroding their bargaining power and straining the relationship between labour and labour-based governing parties. Despite a seemingly universal trend towards a 'loosening' of party-labour alliances, there exists considerable variation between the three cases.

In this chapter we analyse the roots of these traditions and show how they help explain the substantial differences that exist across the three cases with respect to the ways in which workers, union leaders and political parties have responded to the challenge of liberalization.

Labour and Liberation in South Africa

Economic liberalization in South Africa coincided with the political democratization of society and a government committed to address the legacy of racial and economic inequality. Central to the challenge facing the new government was a labour regime shaped by a racially segmented labour market where white workers enjoyed power and privilege and black workers were excluded from skilled and supervisory positions. This structure – what has been called the *apartheid workplace regime* (Von Holdt 2003) – led to waves of worker resistance and organization throughout the twentieth century. The state responded to this worker militancy with repression, culminating in the establishment of the apartheid state in 1948. It is possible to identify three different political traditions within the labour movement which have historically shaped the different perspectives on its relationship with the national liberation struggle. The first, and most powerful, is the national, democratic political tradition.

During the 1950s the African National Congress (ANC), in alliance with the South African Congress of Trade Unions (SACTU), established its leadership among the oppressed classes. This was achieved through mobilizing the oppressed – across class lines – around the demands of the Freedom Charter adopted in 1955. SACTU's participation in the Congress Alliance facilitated the rapid growth of trade unions in certain regions. But it also brought SACTU into direct conflict with the apartheid state and the organization felt the full force of repression in the 1960s. By 1964 it had ceased public activities in South Africa and operated from exile.

In the late 1970s this 'national-democratic' tradition re-emerged in the labour movement with the advent of general unions, particularly in the South African Allied Workers Union, which followed in the tradition of SACTU. These 'community unions' argued that workers' struggle in factories and townships was indivisible, and that unions had an obligation to take up community issues. It was branded 'economism' and 'workerism' for unions to restrict activities to factory struggles.

This 'national-democratic tradition' involved a view that South Africa could not be understood in simple class terms. Social reality was based on a 'colonialism of a special type' necessitating national-democratic rather that class struggle as the appropriate strategic response. This meant a multi-class alliance under the leadership of the ANC, drawing all the sectors of the oppressed black masses and sympathetic whites, and aiming to establish a 'national democracy'. Supporters of this view differed over whether it was necessary to pass through a national-democratic stage before a socialist stage (the two-stage notion of revolution), or whether national and class struggle take place coterminously.

During the 1970s an alternative political tradition developed in the union movement. The 'shop floor' unions that first emerged in 1973 eschewed political action outside production. They believed it was important to avoid the path taken by SACTU in the 1960s. Rejecting the 'community unions' as 'populist', the shop floor unions developed a cautious policy towards involvement in broader political struggles. These unions (particularly those which were to affiliate to the Federation of South African Trade Unions) emphasized instead the building of democratic shop floor structures around the principle of worker control, accountability and mandating of worker representatives. They saw this as the basis for developing a working-class leadership in the factories. These unions argued that this strategy involved the best means of survival in the face of state repression, and of building strong industrial unions, democratically controlled by workers. Some within this political tradition supported the creation of a mass-based working-class party as an alternative to the South African Communist Party (SACP).

The picture of unions and politics in South Africa has often been incomplete through neglect of a third political tradition – that of black consciousness. Its origins lie partly in the Africanist ideology articulated by the Pan-African Congress (PAC), which broke away from the ANC in 1959 because of the latter's 'multi-racial' definition of the nation. Black consciousness has similarities to the national-democratic position, in that it holds that racial oppression is a manifestation of national oppression. However, its emphasis on racial structures and identities virtually excludes class relations from its analysis. This has often given rise to voluntaristic and romantic forms of organization and mobilization.

The formation of COSATU in December 1985 brought together unions from all three political traditions: the well-organized industrial unions drawn from the shop floor tradition; the general unions drawn from the national-democratic tradition; and the National Union of Mineworkers (NUM), having recently broken from the black consciousness tradition.

Responding to the euphoric mood, a COSATU delegation visited Lusaka in February 1986 to speak to the exiled ANC. A joint communiqué was issued in which the independence of COSATU was acknowledged, while the federation committed itself to the struggle for a non-racial South Africa under the leadership of the ANC. This clear identification with the national-democratic tradition brought the legacy of division in the labour movement to the surface again.

Critics within the shop floor tradition believed that the COSATU leadership had acted without a proper mandate in visiting Lusaka and should have made it clearer that COSATU would struggle 'independently under its own leadership' within the broad alliance. By implying that COSATU was 'operating under the leadership of the ANC', critics argued, a spirit of sectarianism could develop towards alternative political traditions. Some within the shop floor tradition went further and argued that the national-democratic tradition stood in absolute contradiction to working-class politics. Organizational style and political content were such, it was argued, that any involvement of the working class in such politics must lead to the surrendering of trade union independence and with it the abandonment of working-class politics.

The political differences that divided COSATU in its first 18 months of existence narrowed and in some cases were buried in the face of the state onslaught during 1987 and 1988. The necessity for unity forced a tactical and strategic compromise, but it had not removed the differences underlying the competing political traditions. Implicit in this broad-front strategy was the view that class contradictions are secondary to the national democratic struggle (Webster & Fine 1989).

In 1989 the reformist F. W. de Klerk replaced P. W. Botha as president and signalled his intention to reform apartheid. In February 1990 he announced the unbanning of the ANC, PAC and the SACP, and freed Nelson Mandela. Over the next 12 months the ANC, COSATU and the SACP forged a formal alliance that cautiously began to distance itself from the armed struggle and insurrectionist elements in its ranks.

It has become conventional wisdom that South Africa had a 'bloodless' transition from apartheid to democracy. This was not the case. More people were killed in the civil war that ravaged KwaZulu-Natal and parts of the former Transvaal from 1990 to 1994 than all the years before that. It began in 1986 in Natal when COSATU members were killed at the Dunlop tyre factory in Howick. Much of this conflict was fuelled by agents from the security establishment of the apartheid state and vigilante groups linked to Inkatha, the Zulu nationalist group and its associated formations. This had a profound impact on KwaZulu-Natal, including Ezakheni, in polarizing civil society between 'comrades' and 'sell-outs'. Trade unions played a key role in negotiating the peace accord signed in 1991, creating a framework for peace to be negotiated at the local level and in bringing about reconciliation within communities such as Ezakheni. This assisted in keeping negotiations at a national level on track.

In addition to its role in the peace movement, five interventions by COSATU during this period helped shape the democratic transition and the nature of the alliance: the combination of mass action with negotiation helped break the deadlocks at important moments; its involvement in economic policy-making generated new institutions, such as the National Economic Development and Labour Council (NEDLAC) and research into new industrial policies; it contributed to the new constitution, including the right to strike; it was a central political actor, mobilizing support for the ANC during the 1994 election; and it initiated and advocated the Reconstruction and Development Programme (RDP). The RDP originated in an attempt by labour to produce an accord that would tie a newly elected ANC government to a labour-driven development programme. In its final formulation, the RDP envisioned as a first priority 'beginning to meet the basic needs of people: jobs, land, housing, water, electricity, telecommunications, transport, a clean and healthy environment, nutrition, health care and welfare' (ANC 1994).

But class contradictions began to come to the fore as the newly unbanned ANC leadership came under a set of local and international pressures to change its economic thinking. From the 1950s, the ANC was publicly committed to a state-interventionist redistributive strategy, as articulated in its central policy document, the Freedom Charter. This policy was publicly reconfirmed by Nelson Mandela on his release from prison in February 1990. If, however, Mandela had entered prison at a time when nationalization was an article of faith, he was released into a world where self-regulating markets had become the mantra.

From 1990, there were a series of economic policy reversals through which the ANC leadership came to adopt positions increasingly consistent with the neoliberal orthodoxy. By late 1993 the ANC had made a number of concessions on macro-economic policy which were to culminate in 1996 in the Growth, Employment and Redistribution (GEAR) strategy (Webster & Adler 1999: 363–371). GEAR aimed to achieve growth through fiscal-deficit reduction, gradual relaxation of exchange controls, reduction in tariffs, tax reductions to encourage the private sector (and especially direct foreign) investment, and privatization. While maintaining in public discourse that GEAR was an extension of the RDP, and even on some occasions the Freedom Charter, these assurances rang increasingly hollow.

When GEAR was released the alliance partners were angered both by the content – which they then saw for the first time – and because the government asserted that it was 'non-negotiable'. COSATU general-secretary Sam Shilowa publicly criticized GEAR and indicated that these policies could never have emerged from the ANC before the 1994 elections. Agreement could not be reached inside COSATU on how to deal with GEAR: some felt that the policy should be given a chance and the government should not be attacked during a currency crisis; others wanted openly to oppose the policy. The issue threatened to be divisive both within and between the alliance partners and, in the words of a leading alliance figure, the 'people walked away from it' (Webster & Adler 1999: 368). Thus, with painful irony, what began as an accord to bind an ANC government to a left-wing development programme ended up ensnaring both COSATU and the SACP in a neoliberal-inspired macro-economic policy (Webster & Adler 1999: 368).

The failure of GEAR to perform anywhere close to its own expectations for growth and job creation led to growing tension between the ANC and COSATU. In 1998 both President Mandela and Deputy-President Mbeki used high-profile speeches to COSATU's Central Committee and the SACP's congress to rebuke both organizations for questioning the government's economic policy. Mandela accused trade unions of being 'selfish' and 'sectoral', bent on protecting their interests at the expense of the nation (Bramdaw & Louw 2000). Compared to the mass of unemployed living below subsistence, workers organized in trade unions were being regarded in certain circles of government as a privileged labour aristocracy.

As the implementation of the new economic policy began to impact on COSATU members, union leaders became increasingly vocal in their criticisms of government economic policy. For the first time COSATU affiliates publicly discussed an alternative to the ANC to lead the Alliance. At the annual congress in August 2000 of the National Union of Metal Workers of South Africa (NUMSA) a resolution was tabled for the SACP to replace the ANC as leader of the Alliance. After considerable debate the congress rejected this resolution and committed itself to 'the strategic relevance of the ANC-led alliance in the current context of the national democratic revolution' (Bramdaw & Louw 2000). At the same time the South African Municipal Workers' Union (SAMWU) resolved at its annual congress 'that the process of beginning a discussion on an alternative to the alliance should begin' (Gumede 2005). Increasingly, the alliance seemed to be no more than an electoral machine that came to life at election time. 'Since the adoption of GEAR by the ruling party', Blade Nzimande, general secretary of the SACP remarked at the SAMWU congress, 'there has been a reluctance within the alliance to meet' (Bramdaw & Louw 2000).

The impact of the ideology of self-regulating markets has not only led to tensions within the alliance; it has also led to an erosion of the organizational base of the union movement through retrenchment and the growing use of various forms of casual employment. This has segmented the labour market into three roughly equal zones: a core of relatively protected workers, a non-core of vulnerable part-time, temporary and contract workers, and a periphery of informal and unemployed workers (Webster & von Holdt 2005). Table 8.1 illustrates the growing informalization of work and unemployment.

The organizational terrain of informal work is very different from that of traditional trade unionism, and workers are turning to alternative forms of organization for support (Buhlungu 2006a). Religion plays a central role in community based organizations, as do relationships between households where food and care-work is shared. This could

Table 8.1 Employment, informal employment and unemployment, 1972–2002

	Formal employment	Informal employment	Unemployment	% unemployment*
1972	6,683,107	432,992	719,124	9.18
1982	8,127,619	1,059,014	737,560	7.43
1992	8,143,003	1,859,990	2,576,027	20.42
2002	7,351,725	3,545,284	4,783,502	30.51

*This is a narrow definition of unemployment which excludes discouraged work seekers; according to the expanded rate of unemployment, the current unemployment rate is about 40 per cent.

Source: Adelzadeh (2003: 238–239)

explain the support among our respondents for controversial, populist ANC leader, Jacob Zuma. As argued elsewhere:

> Their support for Zuma indicated their sense that the trade union movement has been increasingly marginalized within the Alliance, and that its concerns – notably over jobs – have been sidelined in the formation of government policy. From this perspective, Zuma – who since his dismissal (as deputy-president of the country in 2005 for facing corruption charges; subsequently he also faced a rape charge) has been presenting himself as the victim of a conspiracy by shadowy and politically reactionary forces – was adopted by the left-wing of the Alliance as a unifying symbol for the downtrodden, the oppressed and the excluded. He was adept at cultivating this image, which was well-received among a workforce increasingly vulnerable to the insecurities of casualization. (Southall, Webster & Buhlungu 2006: 221)

New social movements, such as the Treatment Action Campaign (TAC) and the Anti-Privatization Forum (APF), are emerging as a powerful voice of these marginalized groups, but whether they will form links with the labour movement and whether, in turn, the labour movement will establish links with them, remains to be seen (Webster & Buhlungu 2004). However, there is a sense in which union leaders are realizing the limitations of their responses to this new wave of social activism and are searching for new responses. As Willie Madisha, president of COSATU, argues:

> COSATU must find ways of pulling new social movements close to COSATU. It must find ways of working closely with them. COSATU, the ANC and the SACP were involved in the formation of the APF, the TAC and the LPM [Landless People's Movement]. Now they have abandoned them and see them as an enemy. (Quoted in Webster & Buhlungu 2004)

On the other hand, labour's central role in the struggle for liberation and the institutionalized voice it has acquired within the policy-making process, enabled it to persuade the Department of Trade and Industry (DTI) in late 2006 to win a temporary reprieve of three years on the importation of clothing and textiles from China. The creation of NEDLAC, as well as its alliance with the governing party, gives labour opportunities to influence policy at the national level in spite of the erosion of its shop floor power base. The result of this weakening at the base and influence at the top is a class stalemate where the forces challenging market liberalization 'do not have the capacity to impose their alternative ideas on either the state or domestic and international capital. Nor is the government and capital able to satisfy the economic demands of this constituency through GEAR. Neither can they crush the opposition' (Webster & Adler 1999: 371).

The reasons for this stalemate will be dealt with in the conclusion of this chapter. What is crucial in this history is that out of the sharp contestation during the 1970s and 1980s a clear democratic form of unionism emerged with a strong emphasis on its autonomy within the national liberation movement. The explanation for the rapid growth of a powerful labour movement lies in its combination of traditional work-place bargaining power with symbolic and moral power that drew on the deep-felt sense of social injustice that underlay the apartheid system.

We turn now to Korea, a country which is also simultaneously liberalizing its economy while attempting to consolidate democracy.

The Legacy of Authoritarianism in South Korea

Although unions have existed in Korea since the nineteenth century, the 'state and employers have considered unions as a danger to social stability and economic growth whenever firms and factories are disturbed by labour disputes' (Song 2002: 198). 'Politicians and employers', Song continues, 'have always maintained that workers involved in industrial disputes are imbued with pro-socialist and Communist ideology. This sort of anti-labour feeling is so deeply rooted in Korean society that organizations such as unions, pursuing social justice and workers' rights, cannot develop normally' (Song 2002: 198). Indeed, at its height in 1989, union density was just less than 20 per cent and has been steadily falling since then to 12 per cent in 2001. A majority of unions remain enterprise-based and there is no centralized collective bargaining (Bae & Cho 2004: 155).[1]

Why have strong labour movements not emerged in the newly industrialized countries of East Asia? This has been the subject of much debate. A common approach, the culturalist argument, is to stress the values of Confucianism, which include loyalty and acquiescence (Clegg, Dunphy & Redding 1986). The idea of labour acquiescence has been tarnished by the emergence of militant labour movements in countries such as Korea. More importantly, cultural traditions have to be maintained and reproduced: they do not operate in a simple way on behaviour patterns. Instead, 'culture' reflects structures; it does not, and cannot, explain them (Wilkinson 1994: 190–195).

A more useful approach has been the 'late development thesis', an argument that stresses the advantages of learning from those who are already industrialized. Influenced by this approach, Deyo concluded that East Asian regimes maintained their power through a combination of repressive measures and an authoritarian corporatist strategy of co-opting most of the relevant institutions, particularly by encouraging enterprise unions and labour-management councils (Deyo 1987, 1989). Deyo notes the importance of timing; by co-opting these institutions before they had a chance to become the basis of independent labour organizations, the elites prevented the formation of a strong labour movement. Alice Amsden provides similar reasons for labour's subordination, focusing on the state's capacity to exert influence over business through the chaebols, companies owned by a small number of politically connected families which enjoy state support for export-led economic development (Amsden 1989).

This strongly anti-left developmental strategy was buttressed by a great deal of ideological and financial support from the United States, who saw South Korea as a key ally in the Cold War.[2] By 1950 South Korea was getting more than $100 million a year from the United States, most of it in the form of outright grants. The entire South Korean national budget for 1951 was $120 million, making the US contribution more than 80 per cent of the total (Cumings 1997: 255).

But the formation of new democratic unions after the four-month 1987 nationwide strike (described in chapter 6), and the active engagement of students and intellectuals in this movement since the 1970s, has led to a reinterpretation of the past and a recognition that workers in Korea were not passive. This has opened up a less pessimistic view of the possibilities of a democratic labour movement emerging in East Asia. Korea is, in the words of Sonn, 'a late bloomer' (Sonn 1997). This can be seen through an examination of its past.

The history of Korea was crucially shaped by its occupation in 1906 – formally annexed in 1910 – by the Japanese, its loss of national sovereignty and the sense of humiliation this evoked among the Korean people (Cumings 1997: 139–184). The emergence of national and communist resistance to Japanese colonialism goes back to the 1920s, when the Korean Communist Party was formed (Cumings 1997: 154–158). The mass mobilization of Korean labour, including the so-called comfort girls, during the Second World War, was deeply resented by the Koreans, a hurt that continues today.[3]

After liberation from Japanese colonial rule in August 1945, those Korean unions which had a socialist vision formed a national centre, the National Centre of Korean Trade Unions (NCKTU), with 500,000 members (Chun 2000: 23). The CIA described these organizations at the time as a 'grassroots independence movement which found expression in the establishment of the People's Committees throughout Korea in August 1945 led by communists who based their right to rule on the resistance to the Japanese' (Cumings 1997: 203). The CIA went on to say: 'The enforced alliance of the police with the Right has been reflected in the cooperation of the police with Rightist youth groups for the purpose of completely suppressing leftist activity. This alignment has had the effect of forcing the Left to operate as an underground organization since it could not effectively compete, in a parliamentary sense, even if it should so desire' (Cumings 1997: 202–203). The result in South Korea was that the US-backed military government and right-wing Korean political groups, with the help of gangsters, crushed these unions and replaced them with an anti-communist trade union movement (Chun 2000: 197–202). This was the origin of the present Federation of Korean Trade Unions (FKTU).

The FKTU was formed to secure the stability of the official union structure under state patronage by deactivating and depoliticizing the newly established unions and transforming them into enterprise-based unions under the administrative and legal control of the Ministry of Labour (Song 2002: 202). Indeed, it could be described as a form of state – or authoritarian – corporatism. However, corporatist control is an internally contradictory mode of incorporating workers because it is vulnerable to rank and file dissent. This discontent was to emerge in 1970 when Korean society was outraged at the self-immolation of Chun Tae-il, who set himself alight after numerous complaints to the Ministry of Labour over the constant violation of labour standards among garment workers (Chu 2003: 227–236).[4]

This was the trigger that was to lay the foundations of the Korean democratic labour movement. Two weeks after Chun Tae-il's self-immolation, on 27 November 1970, the Chonggye Garment Workers Union was formed by women working in the garment and textile district of central Seoul (Chun 2000: 290). Chun Tae-il's death has become a symbol of worker resistance and an inspiration to trade unionists. Using similar tactics to those in Durban at the same time, union activists organized small groups of workers into the union at factory level. This process was known as the Acacia Meeting (Chun 2000: 300). So effective was this method that by 1973 union membership had grown to over 8,000 (Chun 2000: 300). In a convincing challenge to the conventional view, Chun argues that 'the roots of the democratic labour movement lie in the refusal by local-level union leaders, who were predominantly female, to accede either to FKTU control or manipulation by capital' (Chun 2000: 49).

Thus, the origins of the current democratic labour movement lie in the 1970s, when independent unions, with the help of students and progressive churches, challenged the compliant male-dominated unions. The two pre-eminent religious organizations that gave support to working people were the Protestant Urban Industrial Mission (UIM) and the Young Catholic Workers' Organization (Chun 2000: 55). As social ills and political repression became worse, Song observes, 'workers gradually turned away from the illusion of government promises and began to engage in class conflict in close association with revolutionary student groups' (Song 2002: 203). By the end of the 1970s there were 5,000 democratic 'local' unions, dropping sharply to just over 2,000 in the early 1980s (Song 2002: 204).

The re-emergence of a democratic labour movement in Korea took place in 1987, a period of mass mobilization and nationwide strikes known as the 'Ulsan Typhoon or hot summer' (Song 2002: 210). In 1986 labour-intellectuals had established the Korean Labour Education Association (KLEA), similar to the Institute of Industrial Education (IIE) formed in Durban at the same time.[5] These intellectuals helped the new unions to organise, bargain and set up their administrative structures. This grouping of intellectuals was especially important in providing educational support, including books and material.[6] The strikes culminated in 1990 in the establishment of the Korean Trade Union Congress (KTUC), independent of the FKTU.[7]

From its inception the KTUC adopted a militant approach to both state and the employers. This new unionism, known as 'democratic unionism', contained three core components. First, it developed in labour-intensive industries such as textiles and electronics, industries characterized by low wages and long hours of work. Second, this new unionism drew, in particular, on young female workers and a new generation of workers less tolerant of economic inequality and political repression (Song 2002: 211). The driving force of the militancy of the new unionism was the youth, who effectively drew on traditional rituals and popular culture such as drumming, local pop music and working-class songs. Third, the new unionism was driven by a radical, socialist vision, what they described as the 'liberation of labour' (Song 2002: 211).

The challenge facing the labour movement was to change dramatically in the 1990s when the state began, under the impact of globalization, to withdraw its support – cheap loans, tax exemptions and tariff rebates – from the chaebols (Song 2002: 215). This precipitated large-scale restructuring, retrenchment and a phenomenal growth of contingent (casual) employment, reversing the Japanese tradition of lifelong employment.[8] Management attempts to introduce flexibility by amending the labour laws provoked severe conflict between capital and labour, leading to the first nationwide general strike in Korea beginning in December 1996 and continuing until 10 March 1997 (Chang 2002: 22–23).

Drawing on the case of Hyundai, Chang argues that the changing nature of labour control in the aftermath of the economic crisis of 1997 involved three interlocking processes: the so-called flexibility of the labour market on the basis of growing job insecurity and increasing numbers of irregular forms of employment; new human resource management methods on the basis of more competition-based personnel management and a capability based wage system; and the restructuring of workplace organization (Chang 2002: 32–33). These reforms were 'conditionalities' tied to IMF loans to Korea. Our research on LG reveals a similar strategy to that of Hyundai and was discussed in chapter 3.

Although the FKTU still exists and is the largest union grouping, with 900,000 members and 28 industrial-level affiliates, it is increasingly being challenged by the new unions; this has, in recent years, led it to adopt relatively progressive politics. These contradictions are captured in the strikes over the privatization of electricity as discussed in chapter 4. The KCTU had 619,204 members and 16 industrial-level affiliates at the end of 2004 (*Journal of the Contingent Worker Centre*, Seoul). The two federations have begun to cooperate closely in the face of privatization and retrenchments and around their 'common goals of maintaining job security and protecting the rank and file from massive layoffs' (Song 2002: 216).

In 1998, after decades of opposition to government, Kim Dae Jung became president with the support of labour. In order to secure a loan the newly elected government was forced in 1998 to implement the austere fiscal policies determined by the IMF (Shin 2002).[9] To construct a consensus-building mechanism, attempts were made by the government to establish a tripartite labour council in January 1998. In February 1998 the three parties in the tripartite labour council reached agreement around a social compact where labour agreed to the IMF conditions of fiscal austerity and government agreed to the construction of a social safety net and participation in policy-making (Song 2002: 220). However, three days later, at a congress of the KCTU, a majority of delegates rejected this agreement, leading to the withdrawal of the KCTU, although they decided later that year to rejoin the council.

Finally, in 1999, the KCTU congress decided to withdraw from the council again when it became clear to them that the council was ignoring labour's proposals (Song 2002: 221).

Labour's marginalization, and the restructuring that followed, led to an intensification of conflict between the KCTU, on one hand, and employers and the state on the other, leading to violent conflict between workers and the police (Song 2000). Growing disillusionment with President Kim's government led to the formation of the Democratic Labour Party (DLP) in 2000. In 2003, President Roh Moo-Hyun came to power with a strong pro-US policy.

While the election of President Kim Dae Jung had indicated a new stage in the consolidation of democracy in Korea, it did not, Shin argues, guarantee the democratization of industrial relations. 'As neoliberal globalization by the democratically elected government proceeds', he continues 'workers' rights have been severely curtailed and threatened. Contradictory outcomes from democratization and globalization generates the contentious labour politics in which organized labour is more militant and combatant than before and the government policy decreases the possibility of compromise and negotiation' (Shin 2002: 8).

The Korean labour movement played a central role in the transition to democracy but, as Song observes, 'organized labour actually gained much less than it contributed to democratization' (Song 2002: 209). Its high point of militancy was in the late 1980s; since then its membership has declined as it still faces state repression, a hostile public opinion and, in the case of KCTU, financial constraints. Above all, the shift to market-driven politics has changed the nature of the employment relationship, reducing job security as well as lifelong employment. While some unions are beginning to explore new resources of solidarity such as the new social movements, labour has been put on the defensive (Song 2002: 232–233, 236).

The crucial importance of this account of the history of labour in Korea is that, in spite of sharp state repression, a strong, democratic labour movement emerged. Similar to South Africa, it had close links with progressive intellectuals who provided it with an alternative vision of working-class politics.

The Erosion of a Class Compromise in Australia

Similar to South Africa, the roots of contemporary Australia lie in settler colonialism, a process of colonial conquest and the dispossession of the land of the indigenous people. Unlike South Africa, where the settler population came to rely on the labour of an indigenous working class, in Australia the rapidly expanding settler population, built around the 'white Australia policy', soon outnumbered the indigenous Aboriginal population. The result was an economically marginalized Aboriginal population, dispossessed and socially excluded, and unable to establish their presence within the economy or the state.

At the turn of the nineteenth century, the white Australian state was founded on a belief in egalitarianism secured through a method of judicial determination in centralized wage-fixing and the conscious protection of industry and jobs. Social reformism was a response to the great wave of strikes of the 1890s. Recognition of the role and rights of trade unionism and protection of their power to bargain collectively were a central feature of the reforms, and the interventions came to reflect the social identity of the new federal state formed in 1901.

The industrial relations system that was to play a central role in making Australia 'the social lighthouse of the world' was founded on the Conciliation and Arbitration Act of 1904. Trade unions were a recognized, integral component of this system and came to play a key role in securing uninterrupted improvement in wages and working conditions. Between 1939 and 1974, workers' real wages rose by an annual average of 2 per cent, and the 40-hour working week was common by the late 1930s (Kelly 1992: 2; Ward 1958).

These material gains, which were the product of trade union recognition, arbitrated wages and industry protection, represented a century-long, stable and durable class compromise that was the foundation of Australian social democracy. This interventionism was seen to assert the positive social value of egalitarianism. Wage justice through arbitration became the means which marked 'the beginning of a new phase of civilization'. A regulated minimum-wage structure, grounded in the principle of human need and welfare, and not profit and productivity, was enshrined through the famous Harvester judgement. In this, the state set minimum wages by judicial decree rather than abandoning workers to the turbulent waters of market forces. The notion of the cost of living became the basis for wage movements. The president of the Arbitration Commission, Justice Higgins, was ruthless in his enforcement of these principles. In a 1909 case involving one of the large mining companies, Broken Hill Proprietary (BHP), he argued that if the company could not pay the minimum rate it was preferable that it shut down. He commented, 'If it is a calamity that this historic mine should close down, it would still be a greater calamity that men should be underfed or degraded' (Rickard 1984: 173–175). Hence the industrial relations system came to be viewed as 'the greatest institutional monument to Australian egalitarianism' (Kelly 1992: 9).

The compromise has had some degree of cultural impact on class relations in Australia. In certain literature, notions of egalitarianism were presented as a defining feature of Australian society. 'The typical Australian', writes Ward, 'is a practical man, rough and ready in his manners ... He believes that Jack is not only as good as his master but at least in principle probably a great deal better' (Ward 1958: 16).

In conducting this social experiment, Australia was advanced as 'a nation that had become the model social democracy of the world'. The nation had set great store both on the capacity of a state to steer the private economy and, equally, on its capacity to shape and shelter Australia's own distinctive social democratic 'labourism' in the face of external pressures (Pusey 1991: 2). This commitment changed during the 1980s. When labour came to power in 1983, constructive engagement with the global economy was fully embraced. This carried with it the liberal economic assumptions regarding the role of market forces and the state that had shaped the global change. Hence, labour was forced to rethink the notion of 'steering' the private economy. The new government accepted the arguments, advanced by key agencies of the state, that it was in the nation's interest to embrace economic deregulation (Pusey 1991: 2). The deregulation of financial markets, the dollar float, and sweeping reductions in tariff protection were advanced as an answer to Australia's declining competitiveness, growing external debt, and a reliance on commodity exports that were falling in value.

Labour's strategists were of the view that this efficiency drive was the most effective defence of the social democratic project in the new circumstances. The creation of a dynamic, globally orientated economy would generate the wealth necessary to sustain these social commitments. Labour aimed at achieving the best in both spheres. The establishment of a social accord between the Labour government and the union movement was to be the principal mechanism for the realization of this balance. The accord established a process within which the terms and conditions of economic liberalization could be bargained over. In seeking the best possible advantage from liberalization, the accord also sought to shift strategic thinking towards a 'high-road' development path. The corporate sector and unions would join forces to implement advanced manufacturing technologies, skill formation, and the transformation of work organization (Mathews 1989).

Australian Council of Trade Unions (ACTU) strategists promoted what they regarded as a new form of unionism appropriate to this phase of economic liberalization. The emergence of best-practice unionism – or strategic unionism – was regarded as a 'historic watershed', not just for the future direction of the Australian movement but for trade unionism more generally. In 1987 the Australian unions adopted this new role when the leadership accepted the proposals in the document 'Australia Reconstructed' (Australia Reconstructed 1987). This role dovetailed with unending waves of economic liberalization and competition policy of the past decade.

Unions could contribute constructively to this efficiency-drive through the adoption of best-practice unionism, which forged a new strategic relationship with management. Unions were to move beyond a traditional defensive wages-and-conditions strategy to a new and innovative role – 'that of making enterprises more efficient' (Ogden 1993: 32–34). A cardinal feature of the change was a transformation of the traditional relationship between management and unions. 'The ingrained distrust of employers by unions must give way to a recognition that some points of agreement are critical to all of us – for example, the efficiency of enterprises' (Ogden 1993: 59). Workplaces where such a strategic unionism is present 'tend to be more productive' (Ogden 1993: 10). Furthermore, such an orientation produces 'a cooperative industrial relations climate' in which 'unions and management can agree on strategic objectives and negotiate through their implementation. By minimizing traditional conflict but remaining independent, the union can concentrate on wealth creation as well as distribution.'

These strategic shifts by the new Labour government and the ACTU represented a radical policy change as the nation was propelled over a short time span from high protectionism to extreme openness to global forces. In this context, the ACTU policy transformation was as far-reaching. A union movement, noted for its long history of rank-and-file militancy, was now the advocate of a cooperative approach.

The Australian case reveals the inherent risks in this process. Rather than strengthening class compromise, the positive cooperation of Australian unions played a significant role in its erosion. A decade after unions adopted this strategy, research reveals the impact of the change on the material interests of workers. During this period, capitalists advanced their material interests, while those of the working class were eroded.

During the accord (1984–1993), real wages fell, while profits rose. The wages share of non-farm output fell from 69 per cent to 65.5 per cent during the life of the accord, while profits increased from 31 per cent to 34.4 per cent (Ewer et al. 1991). Taking into account both social and real wages, average living standards declined by 5.4 per cent. The pattern of wage restraint was unevenly distributed (Ewer et al. 1991: 32). Those in relatively unskilled work suffered a massive cut of 15.6 per cent. The restructuring has been accompanied by a rise in the level of part-time employment, which has increased at three and a half times the rate of full-time employment. There has also been a notable shift to precarious employment through the increased casualization of work and increases in the proportion of temporary jobs, outsourcing and the use of labour hire companies as enterprises introduced 'numerical flexibility' into the workforce (Ewer et al. 1991: 148).

By 1990 it was estimated that about 25 per cent of the Australian workforce had been casualized (Ewer et al. 1991: 139). Over the decade of global engagement, nearly all industries realized a significant growth in the proportion of casuals. This change affected both men and women. Among men, the proportion of casuals has doubled, rising from 10 per cent to 21 per cent between 1984 and 1987. For women, there was also an increase, but of a lesser magnitude – from 26 per cent to 32 per cent (Ewer et al. 1991: 140). The growth rate in casualization stabilized in 1998 and is currently at 27.9 per cent of all Australian workers; that is one in four (McNamara 2006).

As McNamara points out:

> There is a growing body of research indicating that the job insecurity created by casual working arrangements is associated with adverse outcomes such as increased fatalities, injuries and illnesses; increased exposure to occupational violence and psychological distress; inferior knowledge/compliance with OHS entitlements, standards and regulations; decreased reporting propensity; less or minimum safety training; and reduced occupational training. (NcNamara 2006: 13)

The union movement's membership decline is one of the largest in the world. This is the reality for a movement that had once enjoyed the highest density in the world. In the space of two decades (1976 to 1996), union density dropped by two-fifths, declining from 51 per cent in 1976 to 30.3 per cent in 1997 (Peetz 1998: 7). Thus, a nation that moved swiftly to become one of the world's most liberal economies is also the nation with the largest decline in union membership. Since union power is a critical ingredient in sustaining class compromise, the declining power of Australian unions is significant.

The final stage in the dismantling of social democracy and the construction of a society based on pure market logic was the introduction of the New Work Choices Legislation in 2005. The centrepiece of the conservative Howard government agenda since 1996 is a shift from collective to individual agreements which are known as Australian Workplace Agreements (AWAs). These were first introduced in the legislative changes of 1996 in the Workplace Relations Act. However, over the past decade AWAs were in a minority compared to collective agreements. The system did not take off as intended.

This is the context for the government reforming their 1996 laws by introducing a higher degree of coercion into the system. Whereas before new workers could choose either a collective agreement or an individual contract, the fundamental change in the new system is that employers now have the right to offer only an individual agreement, which of course the worker can refuse at the risk of not being offered the job. The assumption is that the employment relationship is an exchange between equals and not one that is embedded in unequal power relations. Hence, trade unions are seen as irrelevant, a position described as 'unitarism' in the industrial relations literature. Those workers on existing collective agreements can remain on these for the duration of the agreement. However, when the agreement comes up for a renewal, they are faced by the same situation as new workers: accept the individual contract or risk your job. David Peetz describes the impact of the individual contract on workers in the mining industry:

> It wants them to adopt the collective ethos of the mine that the corporation decided to put in place, which is to work 'flexibly', 'productively', 'as directed', and according to a pay scheme unilaterally determined by management. If they have any complaints, they can talk to their supervisor, or to his supervisor. But they cannot involve their own collective body, the union. (Peetz 2006: 2)

Apart from the introduction and strengthening of an individualized tier of employment relations, a second major feature of this model is the systematic unmaking of existing collective bargaining procedures. This includes a high degree of restriction on the role of unions in the workplace and the issues that can be contained in a collective agreement. Penalties are imposed on any negotiations involving these prohibited matters. The following issues are prohibited from collective agreements: anything related to facilitating unionism, such as union training, paid union meetings, payroll deductions for union fees; anything in the agreement that spells out a process of conflict resolution; anything that limits individual agreements; and finally anything in the agreement that spells out protections against unfair dismissal. The laws severely curtail the Australian Industrial Relations Commission, which is a government body set up at the turn of the nineteenth century to conciliate and arbitrate industrial disputes and set minimum wage rates. There are also stringent new restrictions on the right to strike.

Finally, and importantly, a major feature of the law is the criminalization of industrial relations. The laws introduce possible criminal prosecution should trade unions or employees breach the new labour laws. Individual workers can face penalties of up to A$60,000 and individual unions and union organizers can similarly be fined. This criminalization of industrial relations is a replica of labour laws that have been introduced in South Korea (Teicher, Lambert & O'Rourke 2006).

The new laws are already creating uncertainty and insecurity in the lives of Australian workers. The case of the chain store Spotlight has received widespread publicity. New workers were required to sign AWAs, which excluded penalty rates (overtime rates) and cut their wages by A$90 a week. Workers who took their annual leave were treated as terminating their collective contracts and were required to sign new agreements on their return. Annette, one of these workers, received the A$90 a week pay cut, which the company compensated by a 2c increase in the basic hourly rate.

So-called liberalization, in this case, clearly implies a state that is severely interventionist. It created 1,500 pages of labour law to bring the 'flexibilization' of the labour market about. Understandably, these changes have triggered a response from the ACTU. This has taken the form of a sophisticated media campaign of television adverts that highlight the likely impact of the laws on Australian households and their leisure-time activities, such as the father who cannot attend his children's sporting functions. A second dimension of the campaign centres on the 2007 federal election. The ACTU has employed people to work in marginal seats in preparation for the ALP's campaign. The leader of the ALP opposition has stated that, if elected, they will tear up the legislation. The unions have also organized mass rallies to promote awareness of the new laws. Thus far the campaign has not embraced militant protest action and movement politics to force change as occurred during the 1998 maritime dispute, when the Maritime Union of Australia (MAU) used community links and militant action to defeat attempts at introducing labour brokers into the port.

The significance of this account is that a different tradition emerged in Australia, labourism, where the labour movement focuses on the workplace and politics are left to the ALP. Unlike South Africa and

Korea, there is not a clearly identified movement with links to the community, what we have called social movement unionism, in Australia. However, in the face of the attack on labour, activists have begun to reconnect with the community through the Union Solidarity Network. This is an initiative of local union members meeting with community organizations, including faith-based and neighbourhood organizations, to raise workplace issues and their impact on the household and community. We will return to this initiative in chapter 10.

Explaining Difference

In this chapter we have shown how market societies are the outcome of an interaction between two tendencies: the drive to marketization and society's attempt to protect itself from the excesses of this drive. We have seen how societies attempt to protect themselves either by retreating from the market or by mobilizing against the market. These different responses can be explained, we have suggested, by the different labour markets and the different political traditions in the trade unions of these countries.

Australia created the world's first welfare state and this led to a tradition of labourism. This rests on two central tenets: firstly, workers have entrenched rights to collective bargaining and the union confines itself to essentially a collective bargaining role. Secondly, the trade union has a political and parliamentary arm, the Australian Labour Party, which protects workers from cheap, immigrant labour and facilitates local industrialization. With globalization the same political party that created these protections opened the Australian market to unregulated competition. This has led to the rapid decline of the labour movement, widespread casualization of work and disillusionment with the ALP. It has also led to the emergence of a number of political independents, and a possible contestation of the direction of the ALP.

Korea was a late industrializer whose economic success was built on a nexus of trade union repression and co-optation, as well as direct state intervention in the economy. Although a process of democratization began in the late 1980s, labour remains divided and marginal to policy-making. The result is ongoing confrontation between the state and militant sections of the labour movement as workers' bargaining position is weakened by the rapid growth of irregular work. However, at the centre of labour's response is an attempt to organize irregular workers. In order to do this, the KCTU is attempting to rebuild its enterprise unions as industrial unions. We have also seen the emergence of small general unions. A broad coalition of social movements has been set up to address the issue.

Labour's central role in liberation in South Africa secured it key rights over decision making at the national level although, at the same time, these rights began to be eroded at enterprise level as employers successfully by-passed them through the growing informalization of work. This liberalization, in addition to high unemployment, has fragmented workers' organizations and eroded solidarity, compounding South Africa's legacy of social and economic underdevelopment.

We have argued that history matters in explaining the different responses to liberalization in these three countries. These differences emerge sharply when we broaden our comparison to examine the levels of social development in these societies. The Human Development Index (HDI) published by the United Nations Development Programme (UNDP) is the best way of measuring a country's social and economic well-being. It contains three weighted indices: life expectancy, educational attainments and gross domestic product (GDP) per capita. When South Africa, Australia and Korea are compared on these three indices the differences are striking (table 8.2).

Also striking are the differences in levels of social inequality. From table 8.3 it is clear that compared to Korea and Australia, South Africa

Table 8.2   HDI indicators for South Africa, Australia and Korea

	South Africa	Australia	Korea
Life expectancy	47 years	80.5 years	77.3 years
Adult literacy rate	82.4%	100%	98%
GDP per capita	$4,675	$30,331	$20,499

Source: UNDP Human Development Report 2006

Table 8.3   Indicators of inequality, share of income or consumption

	Poorest 10%	Poorest 20%	Richest 20%	Richest 10%
Australia	2.0	5.9	41.3	25.4
South Korea	2.9	7.9	37.5	22.5
South Africa	1.4	3.5	62.2	44.7

Source: UNDP Human Development Report 2006

is a highly unequal society, with the richest 20 per cent of the population accounting for 62.2 per cent of consumption.

If 'development' is defined as increasing the possibilities of more people realizing their potential as human beings, what Amartya Sen (1999) calls 'the capability approach', then South Africa faces the greatest developmental challenge of the three countries. The legacy of colonialism and apartheid has left black communities with a lack of social and economic infrastructure. Indeed, according to the 2003 Human Development Report on South Africa, socio-economic conditions are declining in South Africa; in 1995 South Africa had a HDI index of 0.72 and in 2003 0.67 (Adelzadeh 2003).

The differences between the three countries are most striking when crime statistics are compared. While South Africa has the highest murder rate in the world (1.19538 per 1,000 people), Korea (0.0196336 per 1,000 people) and Australia (0.0150324 per 1,000 people) are dramatically lower and are rated at 38th and 43rd in the world (United Nations Office on Drugs and Crime, Centre for International Crime Prevention, 2003). To put it in simpler language, the murder rate in South Africa is 79.5 times higher than that of Australia and 61 times higher than that of Korea. All three countries have experienced an increase in social problems, but it is most dramatic in KwaZulu-Natal where the HIV/AIDS pandemic has devastated family life. Recent research estimates that 19.8 per cent of families are orphaned (either a mother or father has died) and 3.2 per cent are double orphans (both parents have died) (Shisana et al. 2005: 113). The mayor of the Ladysmith/Ezakheni area estimated that there were 5,000 orphaned families in the area (interview: Duduzile Mazibuko, 2005).

South Africa and Korea differ from Australia in that both societies are going through a 'double transition': attempting to consolidate democracy while simultaneously liberalizing their economies. The process of democratization has opened up new opportunities for labour, while liberalization has eroded its bargaining power. The effect has been contradictory. On the one hand, new rights and institutions have been created, opening up opportunities for labour to move beyond a simple, defensive protest-based politics to what Michelle Williams has called generative politics. 'Generative politics', she suggests, 'is about innovation in collective action that seeks to engender new political actors, organizations and institutions' (Williams 2006: 5).

We have identified examples of these new institutions in both Korea and South Africa. In South Africa the creation of the National Economic Development and Labour Council (NEDLAC) where old – and new – organizations are required by statute to attempt to reach consensus on all significant social and economic policies before they go to parliament, is an example of such an institution at national level. Similarly, institutional innovations emerged at the local level where the Ezakheni/Ladysmith council 'was convening large and extremely lively open-budget meetings at which residents were explicitly invited to educate the councillors about their priorities' (Hart 2002: 254). This led to the establishment of a Local Economic Development Forum 'structured around capital, labour, the local state and NGOs' (Hart 2002: 270).

A similar process of generative politics took place in Korea with the emergence of a militant labour movement and the creation of a tripartite body to deal with the economic crisis in 1997. However, KCTU withdrew from the tripartite body because its voice was not being heard. Indeed, in some cases meetings were taking place without them being involved. The state has succeeded in rolling back labour rights and introduced severe financial penalties from union members or officials who are involved in industrial action.

The most significant conclusion we draw from this history of labour is the way in which the power of global markets has eroded workers' traditional bargaining power. In the case of South Africa the power of African nationalism and the legacy of 'race' has contributed to the erosion of a working-class politics, but what has been decisive is the way in which corporate power has been able to reconfigure the class structure to effectively absorb much of the key leadership of this democratic union movement. Similarly, in Korea the traditional bargaining power of the democratic unions has been severely undermined by a combination of local corporations and the pressure of international financial institutions. The labour movement is at a cross-roads. It needs to identify and draw on new sources of power. Above all, it needs to shift scale and extend its organizational power to the global level to counter that of the global corporation.

In the course of our journey in the three research sites, we discovered such an attempt in Orange where workers embarked on a struggle to globalize their demands. However, this new initiative is a subject of contestation within the union as a division emerged between those who were fatalistic about the prospects of taking on the corporation and simply wished to bargain a decent redundancy package, and those who believed that new strategies were essential to the future of the union movement. This latter position has prevailed because the senior union leadership has endorsed the new strategy and is committed to its implementation. This experiment is yet to fully unfold in Orange.

Is this social form embryonic of a new form of grassroots global unionism or is it a futile and short-lived response to the inevitable logic of global capital? Recently, these new transitional advocacy networks have become the object of serious study (Tarrow 2005; Keck & Sikkink 1998). Their significance has been captured by Appadurai: 'Although the sociology of these emergent social forms – part movement, part networks, part organizations – has yet to be developed, there is a considerable progressive consensus that these forms are the crucibles and institutional instruments of most serious efforts to globalize from below' (Appadurai 2002: 282). Similarly, Castells speaks of two forces in the 'back alleys of society ... In alternative networks [and] ... in grassroots initiatives of communal resistance'. This is where, Castells says, he has 'sensed the embryos of a new society' (Castells 1997: 362).

We turn in the next chapter to an examination of these new forms of transnational trade union activism and focus on the Southern Initiative in Globalization and Trade Union Rights (SIGTUR) in particular.

9

Grounding Labour Internationalism

There is a great debate taking place across the globe, at all levels of the union movement, as to new strategies to confront the power of global corporations that so dominate the economy, politics and society. Like all movement debates, issues are seldom resolved through theorizing alone, but rather through advancing theory through reflection on practical experimentation. We begin this chapter by introducing one such experiment.

In the late 1980s the Australian labour movement faced a dilemma: should it turn inward in the face of economic deregulation and search for national protectionist strategies, or should it establish links with unions struggling for recognition in Asia and search for an international response to the challenge of globalization? A strategy was needed that addressed this new reality – economic deregulation exposed the rights and conditions that unionized Australian workers had won against Asia's cheap, controlled labour – a situation that clearly threatened these historical gains (Lambert 1998: 271; 1999a: 72).

After considerable debate among unionists in West Australia, Unions WA adopted the outward-looking international option. Establishing contact and building enduring relationships with democratic unions in the Asian region became a priority. Individuals who were part of a political generation of committed activists in Asia and South Africa were identified and encouraged to become part of this venture. A research visit was made to the Philippines, Malaysia and Indonesia, where contact was established with struggling organizations facing circumstances similar to those that pertained during the South African labour movement's long and difficult struggle for recognition in the 1970s and 1980s (Lambert & Webster 1988; Lambert 1990: 258; 1997; Seidman 1994).

This social movement unionism (SMU) orientation of the new democratic unions in South Africa resonated with the vision that SIGTUR founders articulated. Indeed, it was felt that given the scale of the challenge unions faced in a deregulated environment the only way to engage was through extending the social movement model beyond the bounds of one nation.

However, SMU was not the only model for establishing transnational links at the time. In the late 1980s, as South Africa was emerging from apartheid, COSATU invited ACTU unionists to share their strategies on globalization. A number of Australian labour-linked researchers and union organizers assisted COSATU unions in developing their response to the global economy. In particular, the programme focused on the metal industry, and a transfer of strategies and policies took place between the AMWU and NUMSA (interview: Dinga Sikwebu, 21 November 2005). The programme was to profoundly change NUMSA's orientation towards the new global economy, from one of resistance to an acceptance of the concept of 'progressive competitiveness' whereby labour adapts to global competition by developing new skills, enterprise bargaining and the discourse of post-Fordism. This was influenced by the ACTU strategic unionism model which was developed during the 1980s when the ALP, under trade union leader Bob Hawke, came to power. However, as Karl von Holdt has demonstrated, this bold attempt to transfer strategic unionism failed (Von Holdt 2003). As in Australia, this approach was contested within COSATU. In 1996, NUMSA invited the Canadian Auto Workers to South Africa, but this attempt to transfer a more militant approach also failed.

Clearly, one shoe size does not fit all, and the particular social and political contexts in which union strategies are developed have to be recognized in these attempted transfers to other countries. An effective labour internationalism is one that is grounded in and grows out of distinctive local histories of struggle and working-class cultures. Within a genuine, action-orientated internationalism, solidarity is viewed as a two-way interactive relationship, not a one-way street. Transposing this 'Australian' model into the South African context was particularly inappropriate given the fact that the AMWU quickly distanced itself from strategic unionism in the 1990s when it experienced the full blast of global restructuring. Corporations in Australia viewed the shift as a sign of union weakness and went on ruthlessly to attack unionism in the name of flexibility, questioning the value of unionism in a modern globalizing economy.

Faced with a democratic transition and economic liberalization in the 1990s, the KCTU took a different approach. They reflected on the

COSATU experience of advancing industrial bargaining, studying the federation's pioneering 1997 document *September Commission of Enquiry into the Future of the Unions*, which was meticulously translated into Korean. This sparked major debates within the KCTU, and some ambiguous translations of concepts added to the controversy. For example, the concept of social unionism, which was central to the Commission's findings, when translated into Korean had a completely different meaning. Instead of indicating a commitment to transformation, lost in translation, the term assumed quite the opposite meaning (Gray 2006; Yoon 1998). Nevertheless, the KCTU did not seek to transplant the South African model to its own context, but was looking for strategies that could be analysed, debated and adapted.

However, while reflecting on and debating contrasting national strategies is a notable advantage in constructing the NLI, provided models are not accepted uncritically, new approaches to labour internationalism transcend the simple exchange of experiences and ideas. Confronted by the power of global corporations, the central challenge of the NLI has become debating, developing and implementing global unionism which seeks to develop a counter-power through working space and scale. In this chapter we track this journey by considering past attempts at building a labour internationalism. We then identify the emergence of a new transnational labour activism, focusing on the nature and constraints of these new initiatives. We then analyse current institutional responses, including attempts by the institutions of the old labour internationalism to revitalize themselves, as well as experiments and new networks that are emerging among activists in the new labour internationalism. We identify key challenges faced by these initiatives to construct an NLI. The question raised by this chapter is whether the two types of internationalisms we identify are in competition or are complementary. Furthermore, how can labour internationalism become grounded in the local, place-based experiences we have considered in this book? What are the opportunities and what are the real constraints to working space in this grounded fashion?

Past Labour Internationalisms

The dominance of free market ideas in the mid-nineteenth century, the First Great Transformation, gave rise to the first labour internationalism, when workers in Europe sought to combine to challenge this logic. In the 1860s, leaders of the embryonic labour movement (consisting predominantly of craft unions) soon recognized that social emancipation required an international component. Local struggles had to connect internationally where the economic dynamic of constantly expanding markets was borderless. The birth of worker internationalism was nurtured in fragile circumstances. Initially, the London Trades Council, which was formed in 1860, drew together the skilled, relatively privileged working class rather than the broad mass of English workers. They established links with French workers whose movement had been inhibited by the repression of the Bonapartist regime and the defeats of 1848. Initial contacts between English and French workers arose because of a common nineteenth-century practice where lower-paid workers from the Continent were imported into England to break strikes. A leader of the London Trades Council proposed 'regular and systematic communication between the industrious classes of all countries' as the solution to this practice (Fernbach 1974). The First International (the International Working Men's Association, IWMA), formed in this context in 1864, became a vehicle for promoting such contact and communication.

However, in the decades that followed, this promising birth of labour internationalism failed to gain momentum. When English building workers took strike action against employers' attempt to reduce wages and increase working hours, they recognized that international support from continental workers in France, Germany and Belgium was 'a matter of life or death'. Employers' threats to import strike-breakers swiftly imposed an international awareness (Olle & Schoeller 1984: 142). However, sectoral struggles against employers intensified the struggle for national trade union rights, which were secured in England earlier than in continental Europe. This was the early sign of the potential of a counter-movement. Olle and Schoeller (1984) argue that the disintegration of the First International was not due primarily to political divisions such as arguments over support for the Paris Commune, but rather to the *national* recognition and consolidation of unions. This constitution of unions as national entities appeared to diminish rather than strengthen the need for internationalism. When legal recognition was won, and as national bargaining systems and more protective trade policies emerged, the view that problems could be fought and resolved at national level, without reference to the international context, dominated.

This shift also reinforced the notion, so prevalent in Britain, that unions represent workers' economic interests separate from political movements. Prioritizing the economic over the political confined future developments to a 'rather feeble' internationalism, which has permeated union activism ever since (Olle & Schoeller 1977). This occurred because capital accumulation on a global scale produced an uneven and differentiated pattern of development, creating competing economic interests and a national fracturing of trade union internationalism. Deep political divisions in twentieth-century labour, following the creation of Stalin's Russia, brought complicating and tragic divisions to bear on an already fractured movement. US Cold War interventions in Asian labour and elsewhere created further divisions that shape today's labour map.

In the context of this bleak history, analysts have concluded that globalization's new international division of labour, the Second Great Transformation, deepened the nationally based conflict of material interest in a divided international movement (Haworth & Ramsay 1986; Ramsay 1999; Olle & Schoeller 1977; Thomson & Larson 1978; Press 1989). The movement of jobs from developed to developing nations, where labour is cheap, overrode any sense of worker solidarity (Olle & Schoeller 1977). Jobs and work lost in one country may benefit workers in another and so there is 'no natural affinity between union movements across the developed world/third world divide' (Waterman 1984: 63, 81). In such circumstances, workers in one country would see workers in other countries as their main enemy, rather than uniting to challenge capital.

Ramsay (1999: 193) contends that in the early 1970s, fear of the emerging strategies of transnational corporations led to renewed debate on trade union internationalism. Unions in North America and Europe were concerned over the capacity of corporations to move investment and jobs to areas where labour activism was weaker and wages were lower. These corporations had the economic power to 'influence governments, withstand labour action and otherwise manipulate their environment' (Ramsay 1999: 193). This image reinvigorated union attempts to internationalize their dealings with multinational corporations (MNCs) through multinational collective bargaining (MNCB).

Levinson (1971, 1972, 1974), then general secretary of the International Federation of Chemical and General Workers' Unions, argued that the economic and social damage of MNCs had to be countered by the growth of MNCB to preserve a democratic balance. Such a strategy would develop unions as 'a truly international force', essential for their survival (Levinson 1972: 141). Levinson's strategy continued to reflect the deep-seated divide between the economic and political spheres. He eschewed the political dimension in favour of an emphasis on the practical economic need for international solidarity identified as essential for the development of MNCB. Levinson's 'evolutionary optimism' (Ramsay 1999: 195) was Eurocentric, drawn from a vision of the evolution of industrial democracy and its extension to the international sphere. In practice, MNCB never made a great deal of headway, apart from some weak attempts to establish World Company Councils by the International Metalworkers' Federation (IMF).

Globalization and New Labour Internationalisms

The collapse of communism, which began in 1989 with the fall of the Berlin Wall, the further deepening of global free trade agreements within the WTO architecture, and the creation of new forms of information and communication technology, have accelerated the capacity of capital to globalize. As we showed in Part One, the shift to market-driven politics on a global scale has radically empowered corporations against society. This has drawn large parts of the globe that were previously insulated from global capitalism, such as Central Europe and the former Soviet Union, India, and China, into possible sites for penetration by global corporations and hence the global labour market. Indeed, the effect of liberalization has been to virtually double the size of the global labour market, creating unprecedented opportunities for the exploitation of unprotected labour. This signalled a significant power shift which has undermined the traditional national approach to protecting labour from the effects of commoditization and cast labour movements in industrialized countries into a crisis.

As Thomas Friedman observed:

> In 1985 'the global economic world' comprised North America, Western Europe, Japan, as well as chunks of Latin America, Africa and the countries of East Asia. The total population of this global economic world, taking part in international trade and commerce ... was about 2.5 billion people. By 2000, as a result of the collapse of communism in the Soviet Empire, India's turn from autarky, China's shift to market capitalism, and the population growth all over, the global economic world expanded to encompass 6 billion people. [This meant that] another roughly 1.5 billion new workers entered the global economic labour force. (Friedman 2005: 182)

While this expansion of the global labour market creates opportunities for capital to exploit labour more effectively on a global scale, it has also ironically led to opportunities to transcend past constraints. This poses a cardinal question: what is the source of labour's power under these new circumstances?

Existing labour internationalism is undergoing a process of change in response to these new circumstances and there is an array of experimentation within the existing international trade unions as well as new networks which are emerging. As Silver (2003) has shown, the working class is made, unmade and remade, and new worker movements emerge under new conditions. Cold War divisions have all but dissipated. The technological revolution brought about by globalization can be used to the advantage of activists – email, websites, databases and many other computer applications are being widely used around the world to find, store, analyse and transmit information. Cyberspace communication systems provide the opportunity to create new networks that reconfigure space, coordinate global campaigns and integrate organization across national boundaries. Indeed, the widespread supply chains inherent in global production introduce new vulnerabilities to enterprises, exposing them to international campaigns and the use of logistical power (Ramsay & Blair 1999; Silver 2003). The emergence of global norms of workplace rights, especially the notion that there are certain core labour rights, creates opportunities for pressure to be exerted on multinational companies breaching these rights. Drawing their strength from symbolic power, codes of conduct are re-emerging as responses to these pressures (Sable, O'Rourke & Fung 2000).

These opportunities have attracted the attention of an older, as well as a new generation of scholars (Lee 1997; Moody 1997; Ross 1997; Waterman 1998; Hathaway 2000; Waterman & Wills 2001; Munck 2002; O'Brien 2000). They have begun to signal a new logic at work in the global economy, a logic that argues that international labour standards must be introduced to prevent a 'race to the bottom'. The disruption of the World Trade Organization (WTO) Ministerial Meeting in Seattle in late 1999 publicly highlighted this new logic. The creation of a World Social Forum in 2000 in Porto Alegro, Brazil, and its rapid expansion annually since then provides further evidence of alternative visions of globalization. These developments point to the emergence of an alternative, counter-hegemonic 'globalization from below' that seeks to challenge the forces of neoliberal globalization (Sandbrook 2003).

Importantly, these authors see the new global economy as an opportunity for labour to realize its historic goal of worldwide working-class unity (Waterman & Wills 2001). They argue that the new information technology provides workers across the globe with the opportunity to communicate quickly and cheaply through networks. However, to achieve this goal, Waterman and Wills suggest that labour needs to think of itself once more as a social movement rather than as a mutual benevolent society (Waterman & Wills 2001: 4).

In Munck's (2002) useful overview of this new global social movement, he identifies two distinct phases of labour internationalism, the 'old' and the 'new'. The former has a 'nation-statist' perspective, is based in the established trade unions of the North, and sees labour internationalism as a form of trade union diplomacy. For many trade unionists, labour internationalism is simply an extension of the foreign policies of their respective national governments, a form of 'trade union imperialism' (Munck 2002: 135–153).

Munck (2002: 151–152) sees the successful developmental solidarity given to the emerging trade union movement in apartheid South Africa as a precursor of a 'new labour internationalism' (also see Southall 1996). In contrast, the new form of internationalism 'has moved beyond a conception of transnational collective bargaining, involving a more social movement unionism' (Munck 2002: 154). This broader conception involves coalitions between labour, environmental and social justice interests, as well as alliances with NGOs, women's movements, consumer organizations and community groups. These alliances cut across the boundaries of national/international, production/consumption and labour/community (Munck 2002: 154–173). They take advantage of information technology to communicate instantly, directly and globally around a campaign-style of organization, which targets 'the weak links in capital's chain' (Munck 2002: 162). Above all, it is more genuinely global and promotes the idea that trade and labour rights should be linked – notions that emerged in the so-called social clause debate (Munck 2002: 128–134). Table 9.1 summarizes these differences between the 'old' and the 'new' labour internationalisms.

Table 9.1 Contrasting 'old' and 'new' labour internationalism

Old labour internationalism	New labour internationalism
Career bureaucrats	Political generation of committed activists
Hierarchy and large bureaucracy	Network
Centralization	Decentralization
Restricted debate	Open debate
Diplomatic orientation	Mobilization and campaign orientation
Focus on workplace and trade unions only	Focus on coalition building with new social movements and NGOs
Predominantly established, Northern, male, white workers	Predominantly struggling Southern Afro, Asian and Latino workers

Despite the stark contrast presented in table 9.1, there are attempts to reform the institutions of the old labour internationalism under new conditions.

The 'Old' Labour Internationalism

The International Labour Organization (ILO), founded in 1919, institutionalizes the idea of social partnership between the state, capital and labour. It sets certain minimum labour standards, codified in conventions. These conventions have become known as the core ILO conventions, including those on collective bargaining (Convention 87), freedom of association (Convention 98) and protection against forced labour, child labour and discrimination in the workplace. These conventions set a basis for a universalized language of rights to counter the commoditization of labour. A key problem, however, is the lack of enforceability of these core standards. The legitimacy of these core standards is also undermined by the fact that countries such as the USA refuse to ratify these conventions, and more recently Australia and South Korea have blatantly ignored these principles in their labour law reforms.

In response to globalization the ILO set up in 2002 the World Commission on the Social Dimension of Globalization, which included prominent leaders from politics, trade unions, business and civil society across the world. The commission presented the ILO with an institutional challenge to recall that it has 'an integrated economic and social mandate and the responsibility to evaluate economic policies in the light of their impact on social and labour policies' (ILO 2004: 3). Indeed, the report is a call for a stronger ethical framework:

> The governance of globalization must be based in universally shared values and respect for human rights. Globalization has developed in an ethical vacuum, where market success and failure have tended to become the ultimate standard of behaviour and where the attitude of 'the winner takes all' weakens the fabric of communities and society. (ILO 2004: 2)

The growing informalization of work and the decline of labour in industrialized countries have led the Bureau for Worker Activities (Actrav) of the ILO to develop two related projects to strengthen trade unions' capacity to engage more effectively in debates and campaigns on social and economic policy issues: the Global Union Research Network (GURN) and the Global Labour University (GLU). The latter is an attempt to strengthen the intellectual and strategic capacity of unions and to establish closer links between unions internationally, and between the labour movement and university based intellectuals. It remains to be seen whether this initiative is sustainable and whether it genuinely contributes to union revitalization over time.

Another important factor to consider is how the end of the Cold War impacted on global union federations. The World Federation of Trade Unions (WFTU), which mainly drew its membership from the old communist world, has receded. The International Confederation of Free Trade Unions (ICFTU), its Cold War opponents, was joined by some of the more radical unions of the South, and attempts were made to reform the bureaucratic organization from within. Indeed, some analysts saw in this strategy the possibility of official international trade unionism engaging 'in a process that may result in its own radicalization' (O'Brien 2000: 534). However, in an analysis of the campaign to bind trade agreements to certain minimum labour rights (the social clause campaign), the author concludes that the ICFTU remains 'entrenched in its classic institutional pluralist approach to the world economy.' He continues:

> If anything, by prioritizing just five 'core' labour standards and promoting the neoliberal language of 'flexibility' and 'partnership', the ICFTU actually abandoned its social democratic model in favour of 'global business unionism'. The ICFTU also retained its classic high-level 'dialogue and agreement' diplomatic lobbying approach throughout the campaign. (Hodkinson 2005: 59)

Hodkinson concludes that the ICFTU has undergone a difficult process of modernization, and has partially opened up its decision-making structures to greater internal scrutiny and democratic participation. After the abortive social clause campaign ended in 2002, the ICFTU went through an 18-month millennium review of the international trade union movement's priorities and strategies. The main outcome of this critical self-evaluation was the historic decision to merge the ICFTU with one of its rivals, the Christian-linked World Confederation of Labour (WCL) in 2006. The merged entity is now called the International Trade Union Confederation (ITUC). However, sceptics argue that, beneath the surface, the ICFTU's new symbolic orientation to alliance-building and membership mobilization is 'a largely strategic manoeuvre to cope with its weakened status within both the international corridors of power and the radical contours of the global justice movement' (Hodkinson 2005: 36; see also Jakobsen 2001). It remains to be seen whether attempts at transformation will open up the possibility of the ITUC becoming part of the new labour internationalism or whether it remains embedded in the old.

A key component of the old labour internationalism are the International Trade Secretariats (ITSs) established in the 1890s, which brought together national unions in a given sector. With the growth of the power of multinational corporations the ITSs emerged in the era of globalization as the key players in the international trade union movement in building countervailing power to global corporations. The most influential ITSs were the International Metalworkers Federation (IMF), the International Food and Allied Workers (IUF) and the International Federation of Chemical and Mining Unions (ICF) (Munck 2002: 145). In the case of the struggle for the recognition of democratic unions in South Africa in the 1970s and 1980s, evidence suggests that the most successful campaigns were coordinated by ITSs, using multinational corporations that were present in South Africa to campaign for the recognition of the emerging democratic labour movement (Southall 1996; Bezuidenhout & Southall 2004).

Under the impact of the new internationalism, the ITSs have changed their name first to Global Union Federations (GUFs) and most recently to Global Unions. In the case of the International Chemical, Energy and Mining Workers' Federation (ICEM), for example, they organized their Second World Congress in 1999 around the issue of 'Facing global power: Strategies for trade unions'. In the Southern African context, the regional structures of the global unions are by far the most involved in transnational union campaigns. This attempt to respond to a new labour internationalism has resulted in sharp contestation within the global unions, which has led in some cases to open conflict. Early in 2006 John Maitland, president of ICEM, resigned because he was disillusioned with the way institutionally conservative forces had undermined attempts to experiment with new forms of organization, connecting the local and the global in ways that challenged corporate power (Lambert 2006).

A part of the response to globalization has been regionalization and the formation of global power blocks. Trade unions have attempted to create structures to engage with the institutions created by these processes. An example is the European Trade Union Confederation (ETUC), based in Brussels, as well as the Trade Union Advisory Council (TUAC), which interacts with the OECD in Paris.

The 'New' Internationalism

In June 2005 the vice-president of the KCTU argued: 'Trade unions must review their directions and strategy. We can no longer rely on traditional methods of struggle. We need a more fundamental transformation of ourselves. If we fail in this endeavour, we have no future. We have to experiment if we want to regain our power' (quoted in Lambert 2006). A number of experiments in new forms of labour internationalism that involve new styles of campaigning and new forms of network organization have emerged. Some of these campaigns also involve the institutions of the 'old' labour internationalism. At this point in time the most significant experiments centre on attempts to imagine new workable models of global unionism, as well as the obstacles that stand in the way of its realization. The bold and imaginative Rio Tinto campaign was a path-breaker in this sphere.

Rio Tinto is one of the world's largest private mining corporations, with 60 operations in 40 nations, mining aluminium, copper, coal, uranium, gold, industrial minerals and iron ore. John Maitland, then president of the Australian Construction, Forestry, Mining and Energy Workers Union (CFMEU) and of ICEM, inspired the initiative which took the form of a networked response to the company's attack on union rights and working conditions.

The Rio Global Union Network (RGUN) was formed by the ICEM and was coordinated from California. Cyberspace was used to promote the campaign, to instantaneously communicate union actions, and to keep track of the corporation's responses. ICEM's demands included a commitment to core ILO Conventions that protect worker rights, the negotiation of a global agreement with effective monitoring mechanisms to give effect to these principles, and the resolution of disputes in the light of these principles. A key innovation of the campaign was the formation of a capital committee that coordinated action at Rio's shareholder meetings. A small number of shares were bought to gain access to these meetings, and the corporation was then publicly confronted with evidence of its abuses of worker and human rights. This is another example of the effective use of symbolic or moral power.

As we described in chapter 7, a working model of global unionism that centres on a small, action-orientated World Company Council (WCC), research capacity and facilitating global networking within the corporation is in the process of evolving out of the Orange experience. Interestingly, a similar model is emerging out of other encounters.

For example, UNI Graphical developed a global union response in its struggle with the Canadian corporation Quebecor, which is one of the world's largest commercial printing companies, employing 43,000 workers in 160 plants in 17 countries across the globe. According to Rosenzvaig, while the corporation conceded union rights in Canada, they were 'living a second life outside of Canada where they became a serial killer of unions', punishing any worker who joined the union.[1] UNI Graphical established the equivalent of the revised WCC, which they call a working group; they created a research capacity, and focused on consumer power and shareholders to pressure the corporation. After years of struggle involving action at shareholders meetings and mass rallies under the banner of 'Justice at Quebecor' the company caved in and signed a global agreement conceding union rights. Rosenzvaig argues that in contrast to some other GUFs, UNI views such an agreement, not as an end point, but rather as an opportunity to focus a continuous campaign. The key, she said, 'is the NLI and global solidarity'.

Similar moves are afoot in the metal industry where the IMF is coordinating a campaign against General Motors (GM) which directly challenges the corporation's right to unilateral restructuring in a determined action to end the bidding war between plants. The unions took three to four years to build a network between plants in GM in order to force a European-wide discussion on restructuring. In another action, when Electrolux announced that it planned to close its German operations and relocate production to Eastern Europe, German shop stewards who were active in the WSF organized a European-wide boycott of Electrolux products and sympathy strikes from other Electrolux plants.

A further initiative is the campaign by the US-based Service Employees International Union (SEIU) to organize in ten cities in collaboration with unions in Australia and New Zealand. This union's highly successful Justice for Janitors campaign is widely seen as a model for the revitalization of trade unions in the industrialized North (Milkman & Wong 2001). This organizing model has been adopted by the ACTU as a way of revitalizing the Australian union movement. In 2004 the SEIU passed a global strength resolution and began searching for global partners to organize in the growing global property services industry. Currently, the provision of property services has consolidated into ten multinational companies employing over 1 million workers, mostly involved in contract cleaning and security services (Aguiar & Herod 2006; Lerner 2007).

SIGTUR

We now turn to a focus on SIGTUR, an experiment in new labour internationalism that has endured for 16 years. SIGTUR is a campaign-oriented network of democratic unions grounded in the global South, committed to resisting the negative aspects of globalization and to constructing an alternative paradigm of global economic relations.[2] More recently at its 7th Regional Congress, held in Bangkok in June 2005, SIGTUR committed to working with the established internationals and the GUFs in particular in confronting the apparent powerlessness of the labour movement in the face of the spatial fixes and other forms of restructuring by global corporations. SIGTUR is in a position to contribute to this all-important project because it represents an attempt to 'identify geographic possibilities and strategies through which workers may challenge, outmanoeuvre, and perhaps even beat capital' (Herod 2001a: 17). This is because the network has reconceptualized spatial praxis, thereby creating the potential for new forms of empowerment. This potential has arisen because SIGTUR is producing new spatial relations between democratic unions in the South that are social movement in orientation with long histories of struggle. The network has consciously eschewed engagement with state and company sponsored 'unionism', as this offers little in the struggle against corporate power. The character of the militant tradition of Southern unions has been forged by state repression where social movement was the only form of power available, hence cutting across space in ways that linked this common experience has proven to be an advance in constructing a counter-power. In SIGTUR's first meetings with the ICFTU leadership in Brussels in 1989, they could not envisage a labour internationalism in any form except the established pattern of spatial linkages from the developed North to the South.

SIGTUR is grounded in union organization, with the potential to avoid the obvious pitfalls of pure networks which are often unaccountable to democratic structures. The weaknesses of pure networks stem from their socially disembodied character. Networking may generate protest politics, as happened in Seattle in 1999. However, to translate these significant and possibly defining moments of protest into an effective counter-movement requires grounding in established unions and civil society organizations. This linkage makes it easier to sustain a network, especially if trade unions are democratic and open to change. Unions also provide a firm financial base on which to build a new global movement that integrates these two organizational forms – social organization and networking – into a coherent whole that draws on the respective strengths of both.

The first tentative steps to establish what was to become SIGTUR were taken in May 1991 when a small conference of democratic unions from the Indian Ocean region was organized in Perth by the Western Australian unions, with the support of the ACTU. Given COSATU's experience in the anti-apartheid struggle, the federation had an immediate impact on new union formations in Southeast Asia that were struggling against powerful corporations, backed by authoritarian regimes. The goal of the initiative was to respond creatively and positively to the challenge of globalization:

> Within Australia we are all too aware of the decisive character of the present decade of global restructuring. How are the Australian unions to compete in a region based on ultra-cheap labour power, secured through severe trade union repression? We can restructure until the cows come home, but we will never hold our ground on existing conditions, while levels of exploitation in the region are so high. This is the point of convergence between the current agenda of Australian trade unions, and the unions in the region. There is an objective basis to the merging of our interests with their interests. (SIGTUR 1998: 2)

On the basis of these common objectives – the shared interest in building strong democratic unions – the initiative grew at a modest but steady rate. The process of building a transnational movement has been usefully delineated by Sidney Tarrow. The first event in the process is that of global framing, a way of organizing experience and guiding action. Global framing is 'constructed by movement organizers to attract supporters, signal their attentions, and gain media attention' (Tarrow 2005: 61). In SIGTUR's case this took place in the early 1990s when Australian trade unionists decided to frame their predicament globally.

This event, according to Tarrow, is followed by a process of internalization, 'the migration of international pressures and conflict into domestic politics' (Tarrow 2005: 79). This happened in SIGTUR in 1995 and marked a turning point for the network, when the recently elected Conservative government in Western Australia intensified its anti-union stance. This radicalized the West Australian union movement, generating strong resistance to the proposed changes to labour law. Pressures on the government, including a threat by the Indian and South African members of SIGTUR to initiate a shipping and trade boycott, led the West Australian government to withdraw the proposed legislation.

Two further processes identified by Tarrow connected this domestic struggle around labour law to the international context. The first is that of diffusion, where the forms of domestic contention were transferred from Australia to India when the fourth SIGTUR congress was organized by the Centre of Indian Trade Unions (CITU) in Calcutta in 1997. At this first meeting outside Australia, the dynamic, style and procedures were shaped by the Indian leadership. Banners and posters lined Calcutta's crowded streets and more than 20,000 workers participated in the opening events. CITU organized factory and community visits that revealed the union federation's social base in West Bengal's working class. The meeting demonstrated the vitality and organizational capacity of a union movement that was deeply rooted in a broader social movement.

The Calcutta meeting brought to the fore the devastating effect downsizing, outsourcing and casualization, and the privatization of state assets were having on the conditions of working people in the South. This meeting also highlighted differences in approach to labour standards, captured in the social clause debate. The Australian delegation was surprised to find the Indian delegates arguing that India could not get rid of child labour, as families were locked into it. They were also surprised and disturbed by the fact that there seemed little engagement with occupational and safety standards in the factories that they visited (Murie, interview, 2003). Importantly this also involves what Tarrow (2005: 32) calls a scale shift, where collective action now takes place at a different level than where it began.

However, two other processes identified by Tarrow take place at the international level and have the greatest potential to create transnational social movements. The first is externalization, the vertical projection of domestic claims onto international institutions; the second is coalition building (Tarrow 2005: 32). The clearest example of externalization was the campaign against Rio Tinto where the union sought out allies in the international arena to bring pressure on the company from outside Australia. This process has also been described as the boomerang effect: 'When channels between the state and its domestic actors are blocked, the boomerang pattern of influence characteristic of transnational networks may occur: domestic NGOs bypass their state and directly search out international allies to try to bring pressure on their state from outside' (Keck & Sikkink 1998: 12).

The second process, the formation of transnational coalitions, takes place when common networks among actors with similar claims develop over time (Tarrow 2005: 32). SIGTUR has over the past 16 years

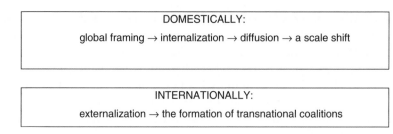

Figure 9.1 The process of building a transnational movement as described by Tarrow (2005)

developed a transnational coalition through its biannual meetings. Although the meetings began in 1991 in Australia, since 1997 when they met in India, they have met outside Australia: in 1999 in Johannesburg, in 2001 in Seoul, and in 2005 in Bangkok, with a planned return to Kerala in India in 2008. Over the past decade, SIGTUR congresses have attracted around 200 delegates from 16 Southern countries.

The Roots of Activism

While Tarrow provides a useful approach to an analysis of the processes involved in the development of transnational activism, he does not provide us with an explanation of why and how someone becomes a transnational activist. Mead's (1934) analysis of how the self-concept emerges provides a basis for exploring this dynamic. Persons achieve the feelings and the idea of a self through seeing themselves through the eyes of the other. Cooley (1956) conveyed this process through 'the looking-glass self' metaphor. Self-image is formed in this process of social interaction where persons take on the mind of the other. The relevance of these insights for commodity status is obvious. When treated as a commodity, feelings of negative self-worth dominate mind and emotion. This status is materialized in poor physical conditions (low wages, long hours, poor health and safety conditions) and in subordination to despotic work relations. In particular, the new regime of flexibility, which for workers means work insecurity, is particularly devastating to a person's sense of self-worth, as they find themselves cast onto the scrapheap. Being treated in this way drains life of meaning and value. Fromm (1947: 72) observed, 'If the vicissitudes of the market are the judges of one's value, the sense of dignity and pride is destroyed.'

Within this predicament, movements create an awareness that this is caused by *social* forces. It is not simply the product of some *personal* inadequacy. This social awareness is a first vital step in the de-commodification of labour. *Participation* in movements creates a counter-experience. Movements depend on the enhanced capacity of individuals for their power. This is achieved by persons discovering and recognizing this simple fact. Their mind, their ideas, their emotion and spirit constitutes the essence of movement and *within* this process of engagement persons experience human liberation *within counter-movements* and can thereby distance themselves from the insidious influence of market ideology. As individuals rediscover their essential worth, their dignity and their inherent value, their sense of anger at the injustice of commodity status is magnified. Such anger is the motor of collective action.

Commodity status is cardinal to the historical experience of the workers who form part of SIGTUR. Their identity is marked by colonialism and racism in which 'coercion – direct and persistent – was an essential element in organizing a labour force' (Munck 1988: 27). Munck (1988: 30) notes, 'Today, the descendents of Indian and Chinese forced labourers are a sizable portion of the working population of Southeast Asia, Africa and parts of the Americas.' This Southern experience was further conditioned by political independence struggles and the disillusionments of the post-colonial era. Our leadership interviews capture this distinctive Southern identity.[3]

These struggles for political freedom occurred at a time of profound transformation in the world economy, what we have called the Second Great Transformation, as the old colonial division of labour gave way to a new international division of labour and the emergence of a substantial manufacturing sector in East and Southeast Asia and now more recently in India. The relocation of production from the developed Northern economies to these newly industrializing societies is a paramount feature of this new economic map, a shift which is being driven by the global policy of self-regulating markets. The new industrialization created a new working class in Asia, youthful and desirous of transcending grinding, rural poverty. However, their aspirations were soon eroded by the harsh conditions of the new factories and they became open to the new democratic unionism that was being advanced in the region (Lambert 1997, 1999b). These are the unions that SIGTUR engaged.

The experience of disenfranchisement and exploitation that characterizes this Southern identity has been exacerbated by the Second Great Transformation. The consciousness of Southern workers is marked by the negative impact of neoliberal globalization.[4] Our surveys of SIGTUR delegates highlighted a common experience of work restructuring in the new era, where insecurity overwhelmed workers as a result of downsizing, casualization and privatization. For those who managed to retain employment, conditions changed with restructuring as they experienced work intensification, deterioration in the quality of their jobs, working and living conditions, reduced real wages and the erosion of culture. The survey's bleak picture of restructuring was reinforced by the view that privatization is leading to a crisis in the reproduction of labour itself.

Significantly, this harsh reality underpinned the delegates' perspective that there is no solution to their predicament through reliance on traditional modes of organizing, hence the stress on the need to build social movements committed to challenging rent evictions, electricity cut-offs, lack of access to public health and poor transportation. Evidence of concerted collective action in the workplace and in the community led delegates to conclude that the most vigorous and determined resistance to neoliberal restructuring had indeed come from the South. Having never experienced the reformist capitalism of the First Great Transformation there is a sense of continuity between the forced labour regimes of colonialism and the restructuring that they now confront. For this reason Southern workers evidence a more overt and determined resistance to restructuring than Northern workers.

Market ideology narrows commitment to individual advancement. Sennet (1998: 30) observes that a market orientation leads to the 'acid erosion of those qualities of character, like loyalty, commitment, purpose, and resolution, which are long-term in nature'. Self-regulating markets on a global scale have produced 'chameleon values', jettisoned as swiftly as a change of clothing as long-term social commitment dissolves before short-term private opportunity. This 'corrosion of character' is the antithesis of social commitment and vision so critical to a counter-movement project.

When SIGTUR was formed the search was for activists with these values, not the time-serving union bureaucrats from company and state-sponsored unions. This choice led to the discovery of a generation of labour leaders who had fought to win democratic union rights in the South, often at a high personal cost. Many in the SIGTUR leadership have endured prison terms, torture and other forms of victimization.[5]

These persons had formed identities centred on the struggle for justice and the empowerment of working people. Interviews reveal how these identities are grounded in value choices: service to the community rather than individual careerism and personal material reward; collective solidarity in place of upward mobility; freedom, democracy and participation over hierarchy and control; social, economic and political equality against elitism; and social control over market logic. This Southern movement has arisen because of the value choices of its leadership. For example, Dita Sari, a leader of democratic unions in Indonesia, chose not to accept the Reebok Human Rights award and an accompanying cheque of US$50,000 in March 2002. The award was to be presented by Robert Redford, Desmond Tutu and other international celebrities at a glittering ceremony planned to coincide with the Winter Olympics at Salt Lake City. In a letter to the Reebok CEO, she stated:

> We know how you treat your workers in the Third World. I know because I helped organize them and carried out actions with them. We know you paid your workers less than a dollar a day when your sneakers were selling for a hundred and that you rented the police to destroy us. Understanding this, we feel that it isn't appropriate for you to put the lid on the wrongs you've committed toward workers by giving this kind of award. (Dwyer 2002)

What is also distinctive about this age cohort is that they were active in a variety of 'social justice' movements in their youth in a similar way to union activists in Southern California (Voss & Sherman 2000). They have carried their earlier vision and organizational experiences into the labour movement. Importantly, they brought to the trade union movement a broader concept of labour that mobilizes workers in the totality of their lives.

Rubina Jamil, president of the All Pakistan Federation of Trade Unions and leader of the Working Women's Organization, illustrates this process. She stressed that it was not possible to recruit women workers into the union movement without a close relationship with the family. Furthermore:

> If there's any problem inside the factory, we try our best to resolve the family problems, too. If the father of the girl needs a job, we try our best, with the help of our male colleagues, to find him a job. If the father or the mother are suffering from any disease we help them to go to the hospital and we assist them, so we are very much in touch with the family members of the women workers and we organize family counselling programmes because we think it is really important for the family members to know why trade unions are important. (Jamil, interview, 2001)

At the centre of the leadership style of these activists is a commitment to union democracy and accountable leadership. A struggle against bureaucratic and corrupt leadership has been at the heart of trade union struggles in Asia in the postwar period. Again, this is captured in the interview with Dass:

> When we took over leadership we posed the question, what is the difference between us and those we kicked out of the union? We did participatory action research. Three to four thousand workers were interviewed. They all said that decision making must be left to the members; the union must be democratic and independent; it should not be affiliated to political parties. (Dass, interview, 2001)

Giving workers a direct voice in organizational decision making builds a positive sense of self and psychologically transforms feelings of worthlessness. Our leadership interviews revealed that the *process* of building democratic unions de-commodified labour. Friendship networks lay at the core of the project. Friendship is about the recognition of uniqueness and the union network is essentially a friendship network, undermining the anonymous, instrumental relationships of the market. As the friendship networks consolidate, a sense of community emerges. These shifts are reinforced by the decentralized network character of these movements, which has the potential to empower new forms of transnational activism.

Constructing a New Labour Internationalism

As we argued in chapter 1, counter-movements do not arise spontaneously, they are constructed. Similarly, a new labour internationalism will have to be constructed. A central obstacle to such a project is that restructuring increases workers' sense of insecurity, turning them inwards. Even when they see the problem as a collective issue, they continue to frame it locally. And when unionists begin to frame the problem globally, they quite often continue to see internationalism as an add-on, rather than a core activity of the labour movement in the era of globalization. This approach to internationalism is illustrated by this observation by a Western Australian union organizer. In an interview he described an incident where he had failed to interest Australian power workers in the attack on Korean power workers. On returning to his office, his union secretary warned:

Don't let your internationalism interfere with your work. You have to bring in the money, organize, and service the members. Internationalism is nice to have, but don't let it get in the way. (Murie, interview, 2003)

As our research in Orange revealed, unionists are attempting to transcend this inward-looking perspective as they search for a practical strategy that frames the problem globally. As an Australian union organizer argued, 'Going global is our only future. We have learned that you have to go international on the big disputes. This is how the world is today' (Fowler, interview, 2003). In a situation where the traditional power of unions has been eroded by global restructuring, the importance of a leadership that is aware of the need to link the local to the global is crucial. Equally important is the need to strengthen the local by deepening links with the community.

Restructuring, we demonstrated in Part Two, is impacting not only on the workplace but is having profound implications for the household and the community. Unions are responding to this challenge by establishing links with the community. An example is the Union Solidarity Network mentioned in chapter 8 that has been established in Australia to link the union to the community. However, instead of asking the community for solidarity in their struggles as unions tend to do, this initiative is an attempt by union organizers to identify and respond to community issues in order to build a genuine union-community alliance.

A case in point is the Pinjara experience in Perth, where union activists had been meeting with local community organizations to assist them in dealing with problems emerging in the community. In November 2006 these community activists supported the AMWU in its dispute with the global corporation Alcoa over subcontracting. The community group held a meeting in the middle of the street, making it a public issue and, by blocking traffic, forced the company to negotiate with the union. This source of power, what we have called logistical power, is a new way of strengthening the local.

These emerging links between the local and the global have been described as a form of 'rooted cosmopolitanism'. Rooted cosmopolitans, Tarrow says, are emerging activists who think globally, but are linked to very real places. In his words:

They move physically and cognitively outside their origins, but they continue to be linked to place, to the social networks that inhabit that space, and to the resources, experiences, and opportunities that place provides them with. (Tarrow 2005: 42)

Figure 9.2 A 'new' labour internationalism

This process is illustrated in figure 9.2.

Linking the union to the community has been at the centre of a broader revitalization strategy in American unions that are attempting to shift unions away from a tradition of business unionism to social movement unionism (Turner & Hurd 2001; Voss & Sherman 2000). Importantly, the leading union in this revitalization process, the SEIU, is also the leading advocate of global unionism in the United States. Under global unionism the world of places is not conceived in terms of bounded spaces. Instead, as Herod suggests, networking scale makes possible place-to-place links across the globe, connecting in a single whole, simultaneously global and local, without being wholly one or the other. This creates a non-hierarchical form of internationalism, organizing locally and connecting place to place globally.

In developing these connections and drawing on these new sources of power, unions have found it necessary to work within traditional union structures, in particular the newly named Global Unions, but they have also found it necessary to go beyond the traditional structures and practices of unions. The issue is not to pit one form of internationalism against another, but rather to examine how existing structures and the new initiatives are building global unionism based on these new sources of power. Such a task will require the identification of new strategies and new forms of worker organization. An example of a new form of organization is the Self Employed Workers Association (SEWA), an innovative response to the growing informalization of work in India (Munck 2002: 111–128).

However, in spite of union resolutions calling on members to broaden their support base to include those engaged in casual work, part-time work and in the informal economy, the majority of unions have failed to make any headway in organizing these vulnerable workers. This challenge will require the formation of alliances between the labour movement and social movements around gender, the environment and other social issues. The union movement will have to place social justice, what we have called moral or symbolic power, as a priority if it is to organize successfully in this domain. Significantly, SEWA has recently been accepted as a member of the ITUC in spite of resistance from local unions still constrained by traditional forms of organization. Indeed, street traders are now being organized internationally through Streetnet International. Similarly, in South Africa, a number of organizations representing informal workers have emerged, most notably the now-defunct Self Employed Women's Union (SEWU) (Webster & Von Holdt 2005: 387–405). Although COSATU has increased its organizing activities among informal taxi drivers, security guards, part-time shop workers, industrial home workers, waste pickers and subcontracted workers, they remain a small proportion of the federation's membership (COSATU 2006, Book 1: 44–48).

The most fundamental challenge to a new labour internationalism remains that of bridging the North-South divide. As Harvey has shown, global corporations are effectively exploiting 'the geography of difference', low-waged zones of the globe, in ways that undermine core labour standards. As a result, claims that a new labour internationalism has arisen in response to the impact of globalization have led to a cautious response by some scholars.

Silver and Arrighi (2001: 530) argue that a North-South divide continues to be the main obstacle 'to the formation of a homogenous world-proletarian condition' and are sceptical of the post-Seattle claims of a New Labour Internationalism based on a Red/Green[6] alliance between Northern and Southern workers. Indeed, Silver (2003: 13) raises the question of 'whether struggles by Northern workers aimed at reforming supranational institutions are more likely to be steps towards the formation of a global working class "for itself," or signs of an emergent, new form of national protectionism'. Seidman (2002) expresses a similar scepticism, highlighting the potential tensions between a language of universal rights and citizenship claims within the nation-state. She articulates the distrust that many developing countries hold of US campaigns for human rights in the Third World, in particular the problem of whether transnational monitoring campaigns empower global managers and consumers in advanced industrial societies, rather than workers in the factories and unions in the South.

Clearly, Seidman is right to emphasize empowering trade unions in the South. As we demonstrate in this chapter, SIGTUR is an example of an initiative to promote trade union rights in the South. But the North-South divide arises out of the uneven development of capitalism from its origins in sixteenth-century Europe, an unevenness that has been exacerbated by the Second Great Transformation. An expression of this divide emerged inside SIGTUR's 6th Congress in Seoul in 2001 during debates on the WTO and how best to respond to terrorism in the wake of the 9/11 tragedy, an event which occurred just six weeks prior to the congress. The differences that emerged in the congress were not over whether the WTO needed to be changed, or whether US foreign policy needed to be condemned; the differences between Australia and the rest of SIGTUR were over how strongly the WTO and the US should be condemned. As discussed earlier, the Southern experience of colonialism and racism, as well as the manipulative tactics of the US towards labour in Asia, has created deep anger, resentment and, at times, visceral anti-American responses. Although these differences initially polarized the congress, a common resolution was eventually formulated and adopted on these contentious issues.

The implications of this incident for constructing a new labour internationalism are crucial: democratic spaces which provide an opportunity for these North-South differences to be openly debated by working people and constructively resolved, need to be consolidated and expanded. However, in the long run, this divide will only be bridged when global institutions are created which move steadily towards even development globally. It is important to stress that globalization is not reducing the poverty gap between the core and the periphery. As Wade demonstrates,

> Over the period of three decades, the per capita incomes of sub-Saharan Africa, Latin America, West Asia and North Africa have all fallen as a fraction of the core's: South Asia's remained more or less constant: East Asia (minus China) rose sharply. But most striking is not the trends, but the sheer size of the percentage gaps, testimony to the failure of 'catch-up'; success-story East Asia still has an average per capita income only 13 per cent of the core. (Wade 2003: 2)

From a movement perspective, these questions of a North-South conflict of interests will not be resolved in the abstract, but rather through struggle itself. Nevertheless, in the shift to pressurizing the placement of the spatial fixes of corporations on the bargaining table there are many obstacles to creating global unionism as an organizational base to achieve this goal. Firstly, there are language barriers, which might be overcome by the smart translation software now available. Secondly, politics needs to be re-evaluated. European trade union politics is dominated by the postwar social compromise, where unions in certain countries have become integrated into corporate structures such as company boards. This mitigates against a commitment to more radical action such as consumer boycotts and pressuring large institutional shareholders to disinvest. Thirdly, unions have to commit to resourcing the construction of global unionism.

However, more positively, the initiatives we have reviewed in this chapter reveal a new engagement with new sources of power and with space and scale to challenge restructuring. Within this process, workers are cooperating across geographic divides, since there is a growing awareness that the interests of workers within the corporation are coalescing in new ways. Corporations now work space relentlessly, moving production to the place of greatest advantage. In this process, all workers experience insecurity, whether they are from the North or the South. SIGTUR receives many requests from Northern unions to engage in a new global union response to restructuring because these unions want the engagement of democratic unions in the South. These new forms of cooperation may herald a new direction in the NLI.

We turn now to the need to develop a vision of an alternative to market-driven development that could become the basis of this common response.

10

The Necessity for Utopian Thinking

We began this journey by telling the stories of three persons: Mpumi Khuzwayo from Ezakheni in South Africa, Bae Hyowon from Changwon in South Korea, and Peter Tyree from Orange in Australia. We also told the stories of their households, their communities and the factories where they work. What Mpumi, Hyowon and Peter shared was that they were all dependent for their survival on the wages they received for working in these factories. While the factories in the three countries had different histories and regimes of control, increasingly they are characterized by market despotism, a model that is grounded in the discourse of neoliberalism. The three workers feared the factories might close down, that they might lose their jobs, or be shifted onto more insecure contracts of employment. These factories did not only manufacture refrigerators, they manufactured insecurity.

In large part workers are responding to this insecurity as a 'personal trouble'. They are adapting to these pressures by working harder, withdrawing into households, or becoming fatalistic. Indeed, in the case of New South Wales, there is shocking evidence of an increase in suicides in response to workplace restructuring (Hunter Taskforce 2001; AMWU 2006). But, as Emile Durkheim demonstrated over a century ago, although suicide is an intensely personal act of despair, it is also a phenomenon that can be explained sociologically. In his pioneering study he showed that acts of suicide are responses to social disintegration – what he called anomie. They can also be a response to deeply felt moral outrage at society's behaviour, what Durkheim called altruistic suicide. This was the case with Chun Tae-il's suicide in Seoul in 1972. Chun Tae-il was making a 'personal trouble' very public, as his recently published diaries movingly illustrate (Chun 2006).

These individual responses are not then only acts of personal despair; they are responses to the pressures of working life, to the shift to what we have called market despotism and the insecurity that this creates. In most cases reactions are expressed in fear, in fatalism, in passivity, in self-blame, but as C. Wright Mills so effectively showed, these personal troubles cannot simply be understood at the individual level. They are the result of the way society is organized. They are, in his words, a public issue.

We have shown how, under the impact of market-driven politics, personal troubles are becoming public issues. Retrenchment, the intensification of work, the growing precariousness of the employment contract, is manufacturing insecurity. This global process of restructuring is not only transforming the world of work, it is also impacting on the households and communities that workers are part of.

In order to understand this process, we went back to the classic work of Karl Polanyi, and his notion of the Great Transformation. We did this because the disruption of society by unregulated markets is not new. It has happened in the past when a similar process of transformation began in the nineteenth century. The danger is that individuals respond to this insecurity in authoritarian ways and draw new boundaries between insiders and outsiders. As Eric Fromm (1947: 86) wrote in the wake of the devastating destruction caused by the rise of fascism in Europe: 'A reaction to acute and chronic anxiety ... is submission to or dependence on an authority.'

A similar phenomenon can be identified in the Second Great Transformation, where communities are fracturing along ethnic lines, with many turning to fundamentalism, whether it be Islamic, Jewish or Christian (Barber 1995; Castells 1997). We have seen 'race' riots in Australia between Muslim Lebanese immigrants and other Australians in the de-industrialized suburbs of Campbell Town, Sydney, the shocking murder of over thirty Somali traders in the suburb of Camps Bay in Cape Town, and the scapegoating of Filipino immigrants in South Korea. Polanyi anticipated and feared these responses to commoditization.

The response of the three communities confirms these reactions, but, reassuringly, it also produced a surprising number of innovative and creative responses to insecurity. We begin by summarizing these responses and drawing out the implications of the different experiments, institutional innovations and global connections at the local level. We do not develop a set of proposals or a programme, but instead identify six key areas that require imagination and hard work if a democratic alternative is to emerge. We then address the feasibility of these ideas by arguing for the necessity of 'utopian' thinking and identify the conditions under which these goals can be realized.

Innovative Local Responses

In Ezakheni old women have got together in poor households in semi-formal community based organizations to generate additional income and provide care for those devastated by the AIDS pandemic. In Changwon, workers have begun to organize among irregular workers, drawing on the traditions of worker organization during the military regime. In Orange, under the influence of the progressive AMWU, activists have begun to address the impact of restructuring by thinking and acting transnationally, moving beyond the trap that there is no alternative. They have begun to imagine a different relationship between society, state and the economy. In doing so they are drawing on the historic role of unions in society.

Unions have historically performed the role of converting 'their organizational space into a political space and contributing to the development of democratic institutions'. By performing this role, 'unions came to occupy a unique position as the purveyors of social cohesion in society' (Jose 2002: 17). This vision of labour is one in which trade unions are not simply concerned with workplace collective bargaining; labour becomes a social and, indeed, a moral actor.

This broader role of labour lies at the centre of its current revitalization strategies. Union organizers are beginning to draw on a neglected source of power, what we have called symbolic or moral power, in the face of the decline in workplace and market bargaining power. To effectively use this source of power, unions have to engage more directly in the public arena, form coalitions with other civil society actors and become part of a broader counter-movement. Importantly, the commoditization of society and the environment affects not only organized workers; it cuts across all strata, including farmers and small business. A facet of union struggle as it emerged historically is the transformation of the insecure commodity status of persons who experience devalued lives into recognition of their creative potential.

Building such a movement in response to the Second Great Transformation is a very different challenge to that facing previous generations. Polanyi wrote at a time when problems could be resolved at the level of the nation-state. Also, there were very real alternatives to unbridled capitalism. The post-1989 world has undermined both

Table 10.1 Local experiments, new institutional forms and global links in Orange, Ezakheni and Changwon

	Local experiments	New institutional forms	Global links
Orange	Attempt at a globalized campaign against closure Coalition with farmers Support for independent political candidate	Linking communities affected by closure within Electrolux Action Committees	Global network in Electrolux
Ezakheni	Historically social movement unionism that linked struggles for land with labour Local government open budget experiment Semi-formal community based organizations – homecare and gardens Government's decision to re-regulate clothing and textile imports	Social movement unionism Open-budget meetings CBO network	Quota on Chinese imports Campaign of the Treatment Action Committee (TAC)
Changwon	Contingency Workers' Centre KCTU organizing clandestinely	General unions Reforming company unions into industrial unions	Successful trade protection strategies

beliefs. It is no longer possible to think that issues can be resolved only at the level of the nation-state and alternatives seem doomed to a distant past. Indeed, neoliberalism has developed a critique of the various alternatives and proclaimed the End of History (Fukuyama 1991). Given the difficulty of this terrain and the fact that it is hard to envisage any dramatic rupture of the present capitalist order, alternatives need to be identified in the actual existing experiments, institutional forms and global links that are emerging (Burawoy 2003). Similarly, John Holloway (2005) has spoken about the 'urgent impossibility of revolution' and the need to accept that 'there can be absolutely no certainty about a happy ending'. Instead, he suggests, we must 'look for hope in the nature of capitalist power itself'. The left, he says, has fetishized the power of the state. Capitalist power is ubiquitous, and so is resistance. Resistance is more partial and incidental than forging collective action; important though this is, it can also be more everyday and routine.

Imagine . . .

Imagine the following scenario relating to an alternative developmental vision of an experiment in Orange: Orange opens up the opportunity for Electrolux workers in the US, Sweden, Germany, China, Mexico, Brazil and Australia to connect through building the currently nominal Electrolux World Company Council (WCC) into a global network.

The WCC has failed to meet regularly because of the costs of setting it up. There are proposals to change these councils into smaller, action-oriented structures in order to encourage, support, coordinate and globally link actions within the same corporation. The primary focus of such a programme would be to challenge unilateral restructuring. Existing worker representative structures in Electrolux are, through the works council system, close to management, and have failed to take up this challenge to date. However, in 2006, German Electrolux workers in Nuremburg, under threat of relocation, linked up with the community in promoting a consumer boycott of Electrolux products. They have also established links, through the works councils, with Electrolux workers in different European countries. This resulted in Italian Electrolux workers going out on strike and demanding that Electrolux shelve its decision to close the plant in Nuremburg.

Imagine in Changwon that the attempts to organize irregular workers succeed and form the basis of a broader social alliance to create a social floor that protects all workers, not simply those who are currently covered by company and private social security. Encouraged by the strike in November 2006, irregular workers throughout Korea succeed in joining progressive unions, who convert company unions into fully-fledged industrial unions, who are better able to globalize their campaigns. Unions in Asia and elsewhere draw inspiration from the Korean labour movement and strengthen their links with the Koreans in order to learn from their experience.

Imagine in Ezakheni that the workers in Defy and other factories form a social alliance with community based organizations and develop an assertive programme to hold the local government to its promises of addressing the issues of crime, disease and local economic development. Workers re-establish their links with movements engaged in land reform and succeed in winning land rights for small farmers. This coincides with a shift in direction by the ANC-led government towards an alternative developmental path, including an unprecedented investment of human and capital resources in education, and industrial policies that redirect investment towards labour-intensive sectors.

Imagine, in all three sites, workers' organizations link up with community and environmental groups in order to articulate a different economic logic – one that sets a social floor and requires companies to abide by strict environmentally sensitive standards. Six components of a new developmental vision emerge from the above discussion.

A new vision of nature

Central to Polanyi's critique of the disembedded market is the destructive impact of these markets on nature, creating the 'wilderness' he refers to (Polanyi 2001: 3). The environmental destructiveness of the market has become the central social issue of the twenty-first century. Joel Kovel describes the market as a 'suicidal regime' and stresses that the contemporary world has forgotten the 1972 manifesto, *The Limits of Growth*. He says we have to rethink our notion of development as economic growth.

> The present world system in effect has had three decades to limit its growth, and it has failed so abjectly that even the idea of limiting growth has been banished from official discourse ... The present system translates 'growth' into increasing wealth for the few and increasing misery for the many ... Growth destabilizes and breaks down the natural ground necessary for human existence. (Kovel 2002: 5)

The problem, as Kovel sees it, is the multinational corporations, which he terms 'ecological destroyers'. 'Capital produces wealth, but also poverty, insecurity and waste as part of its disintegration of eco-systems' (Kovel 2002: 66). The earth is reduced to 'a repository of resources' (Kovel 2002: 125, quoted in Cock 2007: 126).

As Jacklyn Cock (2007) argues, these resources are increasingly likely to become the cause of armed conflict throughout the world. In addition to the obvious conflicts around the control of the supply of oil in Iraq, Michael Clare writes: 'Scarcities of other vital resources, including fresh water, fish stocks, and timber could also provoke fighting between competing states and peoples. Although fresh water is abundant in some (mostly Northern) regions, it is not available in sufficient quality in many areas with large and growing populations – leading, conceivably, to recurring conflict over access to vital sources of supply' (Clare 1998: 73, quoted in Cock 2007: 127).

This inequality between nations is fuelled by the unsustainable consumption patterns by the rich and the exclusion of the poor from livelihoods. At present, Cock writes, about 20 per cent of the world's population consume almost 80 per cent of the world's resources (Cock 2007: 125). 'Poverty alleviation', Wolfgang Sachs writes, 'cannot be separated from wealth alleviation. The affluent have to move towards resource-light styles of consumption' (Sachs 2002: 21). He went on to write in a memo for the World Summit on Sustainable Development in Johannesburg held in 2002:

> The marginalized majority contribute little to environmental degradation. Their per capita use of fossil fuel, water, land, and their production of waste, as well as of greenhouse gasses, is far inferior to the levels maintained by middle and high income groups. The causes of pollution and land scarcity are rather to be found in the consumption patterns of the well-off. The urban poor ... claim little of the resources, but have to bear the bulk of the waste. (Sachs 2002: 31)

To break this logic of 'suicidal' market development, it is necessary to introduce a new economic logic that reconnects social needs and nature. A pertinent example of this iniquity surfaced in our case studies: side by side with the escalating consumer wants of Orange, the majority of working people in Ezakheni lack access to electricity and clean water. It is important to distinguish between the right to certain basic needs such as food, shelter and clothing, with those wants that are constantly manufactured and manipulated by the market.

A new vision of work

The transnational initiatives arising from the restructuring of Electrolux open up the possibility of a new vision of work where security and meaning are manufactured rather than insecurity. There is a growing literature on the new technologies that could create a new balance between work and leisure as well as multi-faceted human development through reduced working hours, production line speeds and the breaking of the work-income nexus. This is captured most imaginatively by Ulrich Beck (2000) where he develops an alternative vision centred on the concept of active citizens, democratically organized in local, and increasingly regional or transnational, networks. Indeed, unions such as the AMWU are already working geographic space to enter into a discourse with other unions in the same economic sector to answer the question of what would be a reasonable line speed that left workers feeling human at the end of a working day.

The majority of the workers in the world are outside standard employment relationships and find themselves in the informal economy. One should also recognize that a 'standard' employment relationship in the industrialized North is the result of centuries of struggles of labour movements, and it is this relationship that is now being eroded. Reducing labour standards in the formal sector, albeit in order to get a foot in the door or to keep a factory where it is, does not fundamentally address the problem.

One way of addressing the problem would be to draw the informal sector into the formal economy, or by empowering this economy in a way that provides an alternative developmental path. Such a path would require a social floor of minimum income and social security benefits. Such a 'levelling-up' of the living standards of working people, especially those in the survival sectors of the informal economy, would significantly increase the market-based bargaining power of working people. This, indeed, would make poverty history!

A new vision of a socially responsible corporation

The aim of such a task will be to socially embed and regulate the corporation, as it is this organization that lies at the centre of market-driven politics. The work of Bakan (2004) suggests a method to engage this critical issue. He argues for a complete revamp of corporate law to widen the definition of the corporation from its narrow bounds of shareholder interests to those of society and the environment. That is to say the state intervenes to ensure that the corporation is harnessed to meet the needs of society.

In recent years, corporations have responded to criticisms such as these by introducing corporate, social and environmental responsibility programmes. However, these so-called 'voluntary initiatives' are often designed to pre-empt the regulation of corporations by governments in the South (Fig 2005). Nevertheless, fair trade organizations have successfully captured the language of corporate responsibility by linking consumer movements to the certification of social and environmental standards in corporations. These schemes draw on symbolic or moral power. Another major emerging source of power is the use of pension funds for what has become known as 'socially responsible investment'.

A vision of an active democratic society

For Polanyi, the expansion of the market threatens society, which reacts by reconstituting itself as active society, thereby harbouring the embryo of a democratic socialism. In other words, if you are trapped in a cycle of work, as in Changwon, or a cycle of poverty, as in Ezakheni, or a cycle of insecurity, as in Orange, you are re-humanized by participation in a movement. A key feature of the active society is that time and space are created for enrichment, where culture, sport and leisure reconnect the individual with nature. This represents the conception of the full all-round development of persons.

We have identified in Ezakheni and Changwon the emergence of social movement unionism. In South Africa this form of unionism emerged in the 1980s, linking workplace issues to the community, or in the case of Ezakheni, linking issues of land dispossession to the labour movement. The democratic form of social movement unionism is a critical source of power, both in the workplace and in civil society, as it creates active agency. This takes participation beyond representative democracy and passive citizenship to new forms of participation that embrace an active civil society. This new dynamic between democratic institutions and civil society facilitates the assertion of a public interest that meets the needs of society (Heller 1999).

A vision of a new fair trade system

A key development emerging from South Africa is the decision by the Department of Trade and Industry to set a quota on the importation of clothing and textile products from China. Before introducing the quota, South African officials visited China and negotiated the issue on the basis of the social impact trade had on employment. The framework through which trade negotiations are conducted in South Africa had been agreed through social dialogue in the Trade and Industry Chamber of NEDLAC. The framework, *South African Trade Policy Framework: Principles and Guidelines for NEDLAC Consultations,* stipulates:

> Trade policy must be constructed at a national level and framed within a broad development strategy that encompasses, amongst other things, policies to limit job loss and promote job creation, reduce inequalities in incomes and wealth, promote local economic development and spatial balance, stabilize the macro-economy, promote appropriate industrialization, strengthen domestic regulatory frameworks, promote education and skills development, and establish social policies. (NEDLAC 2004: 1)

The experience of Japan, South Korea and Taiwan, and now China, demonstrates that countries do not have to adopt liberal trade policies in order to reap large benefits from trade (Wade 2003; Rodrick 1999). All these countries experienced relatively fast growth behind protective barriers. A significant part of these countries' growth came from replacing imports of consumption goods with domestic production, while more and more of their rapidly expanding exports consisted of capital goods and intermediate goods. As they became richer they tended to liberalize their trade – providing the basis for the common misunderstanding that trade liberalization drove their growth.

This is not a clarion call for a return to a past era of tariff protection. It is an attempt to explore the possibilities and limits of regulating trade in a way that brings society back into the economic equation. The South African case is an example of a tentative move towards a socially regulated form of trade. The existing institutions of global governance need to be redefined and reorganized to represent the interests of society. Monbiot (2003) has attempted to imagine the construction of a Fair Trade Organization replacing the World Trade Organization. The goal of these institutions would be to secure even and socially responsible development throughout the world.

A vision of a new global politics

In both Ezakheni and Orange we identified disillusionment with existing political parties. In Korea, an alternative workers' party has been formed. There is clearly a need for an alternative politics that is able to put society back into politics. This is an ambitious task as it requires a reconfiguration of the relationship between the market, state and society. If society consists of institutions inter-positioned between the state and the market, it is these institutions – the schools, the universities, the libraries, the media – that need to be contested and spaces made where markets can be subjected to public control. Building a counter-movement implies developing the structures and social organization that are able to take this contestation forward.

Leys (2001) stresses the need for a new conception of political parties as an alternative to market-driven politics. He shows how existing social-democratic parties, in this instance, the British Labour Party, have been captured by corporations. For us, a new approach to politics would have to link the global and the local. Any single nation-state that attempts to move in the direction of an alternative in any or all of the key areas we have identified would come up against the power of global corporations and global finance. A globally integrated party system will have to retain a high degree of local sovereignty on a range of issues that represent local histories, cultures and societies.

Real Utopias

We have identified six key areas where possibilities exist for an alternative vision of a democratic politics. Discussions on a democratic alternative to the current neoliberal global order often flounder because people believe they are not possible. As Richard Turner wrote in 1972 at the highpoint of apartheid South Africa's repressive regime:

> There are two kinds of 'impossibility': the absolute impossibility, and the 'other things being equal' impossibility. It is absolutely impossible to teach a lion to become a vegetarian. 'Other things being equal' it is impossible for a black person to become a prime minister in South Africa. (Turner 1972: 3)

Indeed, in 1972 in apartheid South Africa, it was not only thought to be impossible, it was treasonous to think along these lines, as Turner sadly discovered when he was banned under the Suppression of Communism Act in 1973. He went on to write:

A social institution is certainly not a solid existing thing like a mountain or an ocean. It may have certain material substrata – written rules and regulations, or a school building to house it in – but ultimately an institution is nothing but a set of behaviour patterns. It is the way in which people behave towards one another. (Turner 1972: 4)

Arguing more recently along similar lines, Eric Olin Wright (2006) suggests that the challenge is to show that alternatives are credible and constitute *real* alternatives. The idea that these structures could be changed in some fundamental way to make life better is, for most people, unthinkable, both because it is difficult to imagine an alternative and also how such an alternative could be realized. So even when our respondents accept the critique of the current world order, their most natural response is to accept it fatalistically as a private trouble, where not much can be done to change it.

The task then of an intellectual, engaged with the logic of disembedded markets, is to develop systematically a conception of a democratic alternative. This is what Wright calls 'envisaging real *utopias*, because of the way the analysis of alternatives embodies the highest aspiration for a world in which everyone has access to the social and material means to live flourishing and fulfilling lives, but *real* utopias rather than fantasies because of the attempt to formulate workable designs for viable institutions' (Wright 2006). Put differently, Turner (1972: 1) wrote that utopian thinking is essential to the evaluation of existing institutions and practices for it is a way of 'exploring the possibilities of a future which could afford as creative and fulfilling a life as possible'. Such thinking constitutes a challenge to all accepted values and an invitation to continuous self-examination.

The crucial point is that the alternative is not given. It has to be imagined and constructed, especially as neoliberal hegemony has knocked the imagination out of the movement. Importantly, such an imagination has to be drawn from distinct places: it has to be grounded in the local, but it also has to connect place to place. Here place and power are being reimagined. This is the challenge facing engaged intellectuals and activists in the era of globalization. A range of transnational activists, in peace movements, in environmental movements, in women's movements, and above all in the World Social Forums, are increasingly framing their activism globally.

It is possible to identify two possible outcomes of these counter-movement activities: the first is an extension of the role of the state to seek a 'societal' version of a social democratic-style arrangement with business, labour, and civil society. The second, a more radical outcome, would be a counter-hegemonic politics that extends the role of civil society and seeks to establish alternative forms of social organization, politics and logics of accumulation.

New Sources of Power

We have spelled out new possible sources of power. The challenge is to identify the points of vulnerability, and how these could help construct a global unionism that challenges the market logic of corporations. The Rio Tinto campaign has shown a point of vulnerability in the capital markets of these corporations. Many are dominated by large institutional investors that depend in no small measure on the pension funds of workers. The Rio Tinto campaign showed that it is possible to pressurize the corporation through an active presence at this level. Hence, capital committees have been established by unions that focus on the share portfolios of companies. The Service Employees International Union (SEIU) is also using this strategy by focusing on the global provision of property services where they have identified €6.9 trillion in the 300 largest pension funds globally.

A second area of vulnerability is the high trade dependency of the global economy. Unions are only just beginning to explore ways in which the movement of goods and services can be disrupted as a mechanism to force corporations to the bargaining table.

A third point of vulnerability is the dependence of corporations on successful marketing of products on global consumer markets. Here, image and brand are all important. In a fiercely competitive market, corporations can lose market share as quickly as they can gain reputation. Hence, they are vulnerable to consumer boycott campaigns. The German workers in Electrolux used this strategy to such good effect that the directors of Electrolux observed that they had rapidly lost market share. The best-known examples are the anti-sweatshop movements in the US against companies in the footwear and clothing industry (Ross 1997).

Conditions for the Emergence of a Counter-Movement

Identifying points of vulnerability in this way is central to developing a counter-movement which strives to assert the needs of society against those of the market. But exploiting vulnerabilities can be counter-productive and lead to alienation of the public unless attention is paid to gaining and retaining the support of the public. While we would agree with Holloway (2005) that taking state power without transforming capitalist social relations does not create 'a society of mutual recognition of dignity', it is also important not to fall for the myth of the powerless state. The capacity of the state remains central to any alternative democratic project (Weiss 1998).

What, then, are the conditions for the emergence of a counter-movement? The starting point is to develop a critique of the existing processes of global restructuring. This requires demonstrating how the self-regulating market manufactures insecurity in the workplace and how these tensions are being transferred to the household and the community, creating what we call a crisis of social reproduction. The second condition is to begin to develop realistic alternatives to the current institutions and social structures, what Wright calls real utopias. This involves formulating workable solutions to these public issues by bringing society back into the forefront of policy-making. The final condition is to map out how to get to this possible future.

We believe we have fulfilled the first condition in the first two parts of this book. Part three begins to grapple with the second condition by identifying what a realistic alternative to market fundamentalism could look like. We briefly address the third condition – how to reach this possible future – but it is a task that cannot be done through scholarly endeavour alone. For such an ambitious project to succeed, it will require a working relationship between intellectuals and these new movements.

This requires a different type of intellectual to the professional, ivory-tower academic reflecting in a book-lined office. Burawoy, rightly, argues that this new type of intellectual should not be 'the legislator of classical Marxism, who would formulate the laws of the collapse of capitalism, or the organic intellectuals of a working-class revolution'. Instead, he suggests, this new public intellectual should be 'an ethnographic archaeologist who seeks out local experiments, new institutional forms, real utopias if you wish, who places them in their context, translates them into a common language, and links them one to another across the globe' (Burawoy 2003).

This conception of the new public intellectual as a global ethnographer is an attractive, even exotic, idea. It captures evocatively the vital role that intellectuals can and should play in the public domain. However, the challenge is to go beyond the notion of a public intellectual to that of an engaged intellectual. For us, although the primary role of the intellectual remains the production of new knowledge, it includes making sure that these ideas impact on those engaged in the construction of a counter-movement.

To be an engaged intellectual, then, is to actively ensure that the knowledge produced becomes part of these movements, through teaching, organizing or publishing in accessible journals and other outlets. But an engaged intellectual is not the same as an organic intellectual. Organic intellectuals operate inside organizations and, while they perform an intellectual function, they do not have the distance or the time to produce the concepts and theories that define the central characteristics of an intellectual. Neither do we see the engaged intellectual as the political vanguard of these movements bringing the 'correct theory' to 'empty-headed and passive workers'. Instead, we see the relationship as a reciprocal one whereby the movements are as much shaped by the ideas of the intellectual as is the intellectual shaped by these movements.

What is at stake here is a division of labour between these inter-related functions. The relationship has the potential for creativity and empowerment when these roles are connected through a shared vision of a democratic alternative. However, equally, the relationship could be one where the autonomy of intellectual work is subject to censorship, or a cynical and instrumental use by intellectuals of these movements for personal gain (see Buhlungu 2006b, 2006c; Maree 2006a, 2006b). The relationship between the engaged intellectual and the workers' movement is a complex and at times contradictory one. University based intellectuals who engage in these movements are in a contradictory location. On the one hand, they have to respond to the needs of the workers' movement; on the other, they are accountable to their peers for their scholarly output. In general this contradiction has been resolved by either immersing oneself within these movements, or retreating into academe. An alternative, more difficult path is to recognize one's contradictory location, build institutional space within the university, contest it on its own terms when it breaches its commitment to open scholarly production, and develop structured reciprocal relations with the workers' movement.

We have now reached our destination. But as with all destinations it is the beginning of another journey. It is a journey on which we invite you, the reader, to join us. It will be a journey where each of us will have to imagine that another world is possible – and one where your help is needed to construct it. The alternative – capitalism without countervailing forces – if history is to repeat itself, will lead us to a grim future indeed. In the 1930s and 1940s in Europe, argues Polanyi, fascism was produced as a result of the failure to govern the social impact of the 'free' market, and the ensuing political and social crises. Today, this is the challenge we face: re-embedding the market in society through expanding social citizenship, or confronting social disintegration and the return of the authoritarian state.

Notes

Chapter 1 The Polanyi Problem and the Problem with Polanyi

1 In the South African liberation struggle, much of the protest was directed at market restrictions, such as pass laws, which restricted job mobility, and restrictions on Africans trading freely in certain areas.
2 There is an urgent need for a research project that is global in scope, which analyses national laws relating to consumer boycotts, the disruption of container movements and communications and information systems. Such a project should explore scope for circumvention.
3 The emergence of new forms of finance capital such as hedge and private equity funds is of enormous consequence for the processes we outline here. Even conservative financial analysts are fearful of the new dynamic of corporate restructuring they have set in train, referring to them as a 'wild west mentality' and 'an accident waiting to happen', where the dominant philosophy of corporate acquisitions is 'buy it, strip it, flip it' (*Australian Financial Review*, 6 March 2007, p. 63).

Part I Markets Against Society

1 More information on *kimchi* is available on the website of the Korea Food Research Institute: www.kimchi.kfri.re.kr/html_en/html/kimchi_01.htm.

Chapter 2 Manufacturing Matters

1 Defy has since been taken over by a Swiss multinational corporation.
2 I. Salgado, 'Malbak unbundling picks up pace', *Business Day*, 30 January 1997; L. Mnyanda, 'Defy bought by local consortium', *Business Day*, 26 February 1997.

3 Ibid.; I. Salgado, 'Malbak sticks with packaging in unbundling scheme', *Business Day*, 7 February 1997.

4 Part of the First National Bank group of companies, the financial group that was set up to take over from Barclays Bank when it withdrew from South Africa following the campaign for sanctions against apartheid.

5 L. Mnyanda, 'Defy bought by local consortium', *Business Day*, 26 February 1997.

6 N. Jenvey, 'Electrification success boosts Defy turnover', *Business Day*, 17 December 1997.

7 Business Day Reporter, 'Defy Invests R15m to boost productivity', *Business Day*, 23 December 1997.

8 www.lge.com/about/corporate/html/company_overview.jsp (last accessed November 2006).

9 www.lge.com/about/corporate/html/company_vision.jsp (last accessed November 2006).

10 The Anglo American Corporation ranks 103rd, with a market value of $58,675.90m.

11 SAB Miller ranks 230th, with a market value of $29,508.10m.

12 Richemont ranks 283rd, with a market value of $24,965.10m.

13 BHP Billiton has a dual listing in Australia and the United Kingdom. The corporation is the result of a merger between a South African mining company that moved its primary listing to the London stock exchange and an Australian counterpart.

14 Rio Tinto also has a dual listing in Australia and the United Kingdom.

Chapter 3 The Return of Market Despotism

1 In South Korea we took into account the peculiar industrial structure of LG's local operation when we considered our sample and included workers from its component suppliers. In Changwon, twelve of our respondents worked as assembly workers. A further four were supervisors, three were involved in quality control as inspectors, and one was employed as a clerical worker. In Ezakheni, ten were employed as assembly line workers, four worked as quality controllers or inspectors, and two were 'repairmen'. Three more workers were interviewed: a worker in the stores, a spray painter and a welder. In Orange, nine of our interviewees worked as process workers – i.e. as assembly line workers. A further four workers were employed as a 'storeman', a foamer, a fitter and turner, and a welder.

2 Worker interview 1-1; Changwon, July 2005.

3 Worker interview 1-3; Changwon, July 2005.

4 Worker interview 1-4; Changwon, July 2005.

5 Worker interview 1-2; Changwon, July 2005.

6 Worker interview 2-1; Changwon, July 2005.

7 Worker interview 2-2; Changwon, July 2005.

8 Worker interview 2-4; Changwon, July 2005.

9 Worker interviews 3-2, 3-3 & 3-4; Changwon, July 2005.

10 Worker interview 3-1; Changwon, July 2005.

11 Worker interview 4-4; Changwon, July 2005.

12 Worker interview 4-2; Changwon, July 2005.

13 Worker interview 4-4; Changwon, July 2005.

14 Worker interview 4-1; Changwon, July 2005.

15 Worker interview 4-3; Changwon, July 2005.

16 Worker interview 1-1; Changwon, July 2005.

17 Worker interview 1-3; Changwon, July 2005.

18 Worker interview 1-4; Changwon, July 2005.

19 Worker interview 1-2; Changwon, July 2005.

20 Worker interview 2-1; Changwon, July 2005.

21 Worker interview 2-4; Changwon, July 2005.

22 Worker interview 3-2; Changwon, July 2005.

23 Worker interview 3-4; Changwon, July 2005.

24 Worker interview 3-2; Changwon, July 2005.

25 Worker interview 4-1; Changwon, July 2005.

26 Worker interview 4-2; Changwon, July 2005.

27 Interview with Leon Adrewartha, the Director of Manufacturing for Electrolux Australia, Orange, August 2003.

28 Interview, Orange, June 2004.

29 Interview, union delegate, September 2003.

30 Interview, 26 April, 2005. This viewpoint was expressed at some point in all the interviews, revealing a common experience of how markedly different the current work regime is from the earlier period.

31 Worker interview 11, Orange, September 2005.

32 Worker interview 5, Orange, September 2005.

33 Worker interview 9, Orange, September 2005.

34 Soon after their election in 1996, the government passed the Workplace Relations Act, which promoted a system of individual contracts (Australian Workplace Agreements), while imposing severe restrictions on unionism. The law changed again in 2005, which expands these restrictions (Sadler & Fagan 2004: 31–33).

35 Interview, Orange, June 2004.

36 Interview, union delegate, September 2003.

37 The unions requested that the Australian Industrial Relations Commission conduct the ballot to ensure the veracity of the vote. This was denied and Electrolux was allowed to appoint corporate accountants KPMG to administer the ballot.

38 Interview, AWU organizer, September 2003.

39 Interview, Mike Lindsay, former manager of the Ezakheni industrial park, December 2005, Ladysmith.

40 The Wiehahn Commission of Enquiry was set up in 1977 to consider ways to address labour unrest. It recommended that black workers be allowed to join registered trade unions. This culminated in the Industrial Conciliation Amendment Act of 1979.

41 Worker interview 3, Defy, Ezakheni, June 2001.

42 Worker interview 8, Defy, Ezakheni, June 2001. These 'stages' refer to steps firms have to take in accordance with South African labour law to lay workers off, including seeking alternatives in consultation with workers or their representatives.

43 Worker interview 9, Defy, Ezakheni, June 2001.

44 Worker interview 1, Defy, Ezakheni, November 2005.

45 Worker interview 2, Defy, Ezakheni, November 2005.

46 Worker interview 9, Defy, Ezakheni, June 2001.

47 Worker interview 9, Defy, Ezakheni, June 2001.

48 Worker interview 3, Defy, Ezakheni, November 2005.

49 Worker interview 6, Defy, Ezakheni, June 2001.

50 Worker interview 2, Defy, Ezakheni, June 2001.

51 Worker interview 2, Defy, Ezakheni, June 2001.

52 Worker interviews 1 & 6, Defy, Ezakheni, June 2001.

53 Worker interview 6, Defy, Ezakheni, November 2005.

Part II Society Against Markets

1 This is a term used by Friedman (2005) to describe the competitive edge Chinese companies have in the global market.

2 Although migration is an individual retreat from change in the 'home' site, it can have contradictory implications, both because migrants are quite often part of informal social networks that make collective action easier, and because it can also be a way of transmitting radical ideas across the globe.

Chapter 5 Strong Winds in Ezakheni

1 Quoted from 'Homeless', by Paul Simon and Joseph Shabalala. Parts of the song are based on a traditional Zulu wedding song. Paul Simon, *Graceland*, Warner Brothers Records.

2 Worker interview 1, Defy, Ezakheni, November 2005.

3 Worker interview 11, Defy, Ezakheni, November 2005.

4 Worker interview 10, Defy, Ezakheni, November 2005.

5 Worker interview 7, Defy, Ezakheni, November 2005.

6 Worker interview 6, Defy, Ezakheni, November 2005.

7 Worker interview 16, Defy, Ezakheni, November 2005.

8 Worker interview 9, Defy, Ezakheni, November 2005.

9   Worker interview 3, Defy, Ezakheni, November 2005.
10 Worker interviews 3, 9 and 10, Defy, Ezakheni, November 2005.
11 Worker interviews 2, 4, 5 and 8, Defy, Ezakheni, November 2005. One of the interviewees was registered as a permanent worker the previous year (interview 14).
12 Worker interviews 15, 17 and 19, Defy, Ezakheni, November 2005.
13 Worker interview 1, Defy, Ezakheni, November 2005.
14 Worker interview 7, Defy, Ezakheni, November 2005.
15 Worker interview 10, Defy, Ezakheni, November 2005.
16 Worker interview 11, Defy, Ezakheni, November 2005.
17 Worker interviews 3, 6, 16 and 18, Defy, Ezakheni, November 2005.
18 Worker interview 1, Defy, Ezakheni, November 2005.
19 Worker interview 2, Defy, Ezakheni, November 2005.
20 Worker interview 10, Defy, Ezakheni, November 2005.
21 Worker interviews 12 and 18, Defy, Ezakheni, November 2005.
22 Worker interview 16, Defy, Ezakheni, November 2005.
23 Worker interview 7, Defy, Ezakheni, November 2005.
24 Worker interview 6, Defy, Ezakheni, November 2005.
25 Worker interview 11, Defy, Ezakheni, November 2005.
26 This is not new or peculiar to Ezakheni. Writing in the 1990s, Beittel suggests that 'On the Rand today, it is common to find Black households which derive substantial portions of their income from non-wage sources, as well as White households which are not completely reliant on wages' (1992: 224).
27 Of course, based on such a small sample, one cannot generalize these findings to all workers at Defy. Nevertheless, when one considers the average wage of STCs of R1,943 per month, compared to an average income of R3,395 per month of permanent employees, this extensive gap in income does seem quite stark. One also has to consider that STCs do not have access to non-wage benefits such as a provident fund and medical aid.
28 Worker interview 3, Defy, Ezakheni, November 2005.
29 Worker interviews 6, 9 and 11, Defy, Ezakheni, November 2005.
30 Worker interview 8, Defy, Ezakheni, November 2005.
31 Monthly social grants vary; R850 old age pension; R640 disability grant; R190 child grant for parents with no other means of support. The average HIV infection rate for the KwaZulu-Natal province is estimated to be 34 per cent (Emnambithi Local Government 2002: iv). Those with full-blown AIDS are entitled to the disability grant.
32 A stokvel is a community savings scheme common in the black community; it provides small-scale rotating loans to its members.
33 Inkatha Freedom Party, Mangasotho Buthelezi's Zulu ethno-nationalist party.
34 Worker interview 11, Defy, Ezakheni, November 2005.

35 Worker interview 5, Defy, Ezakheni, November 2005.
36 Worker interview 6, Defy, Ezakheni, November 2005.
37 This is a reference to the civil war that took place in KwaZulu-Natal. This is discussed in chapter 8.
38 Worker interview 8, Defy, Ezakheni, November 2005.
39 Worker interview 3, Defy, Ezakheni, November 2005.
40 Worker interview 9, Defy, Ezakheni, November 2005.
41 Worker interview 10, Defy, Ezakheni, November 2005.
42 Worker interview 17, Defy, Ezakheni, November 2005.
43 Worker interview 14, Defy, Ezakheni, November 2005.
44 Worker interview 11, Defy, Ezakheni, November 2005.
45 Worker interview 19, Defy, Ezakheni, November 2005.
46 Themba Qwabe, interview, 2005.
47 Mbuso Mchunu, interview, 2005.
48 Worker interview 17, Defy, Ezakheni, November 2005.
49 'Khomanani Caring Together' is a R100 million government funded nationwide communications campaign which provides information and a strategic response to AIDS, tuberculosis and STIs (sexually transmitted infections), in general (Department of Health 2003).
50 Elizabeth Hlatswayo, interview, 2005.
51 Worker interview 12, Defy, Ezakheni, November 2005.
52 Worker interview 11, Defy, Ezakheni, November 2005.
53 Worker interview 6, Defy, Ezakheni, November 2005.
54 Worker interview 15, Defy, Ezakheni, November 2005.
55 Worker interview 1, Defy, Ezakheni, November 2005.
56 Worker interview 2, Defy, Ezakheni, November 2005.
57 Worker interview 3, Defy, Ezakheni, November 2005.
58 Interview, Elizabeth Hlatswayo, community worker, 2005.
59 Interview, Elizabeth Hlatswayo, community worker, 2005.
60 Worker interview 5, Defy, Ezakheni, November 2005.
61 Worker interview 17, Defy, Ezakheni, November 2005.
62 Worker interview 11, Defy, Ezakheni, November 2005.
63 Worker interview 4, Defy, Ezakheni, November 2005.
64 Interview, Elizabeth Hlatswayo, community worker, 2005.
65 Worker interview 2, Defy, Ezakheni, November 2005.
66 Interview, Elizabeth Hlatswayo, community worker, 2005.
67 Interview, Elizabeth Hlatswayo, community worker, 2005.
68 Worker interview 6, Defy, Ezakheni, November 2005.
69 Duduzile Mazibuko, interview, 2005.
70 'The mayor of Ladysmith stated that the increase in the cost of water was decided "unilaterally" by the IFP district council, while Mr Lindsay states that the R1.53 per kilolitre at which water used to be charged was the result of incorrect calculations made by the ANC town council' (Fakier 2005: 54).

Chapter 6 Escaping Social Death in Changwon

1 www.changwon.go.kr/foreign/english/ (accessed 16 October 2006).
2 www.changwon.go.kr/foreign/english/ (accessed 16 October 2006).
3 www.changwon.go.kr/foreign/english/ (accessed 16 October 2006).
4 www.redbrick.dcu.ie/~melmoth/korea/changwon.html (accessed 16 October 2006).
5 www.changwon.go.kr/foreign/english/ (accessed 16 October 2006).
6 Worker interview 2-4, Changwon, July 2005.
7 Worker interview 4-2, Changwon, July 2005.
8 Worker interview 4-4, Changwon, July 2005.
9 Worker interview 5-2, Changwon, July 2005.
10 Worker interview 3-1, Changwon, July 2005.
11 Worker interview 1-4, Changwon, July 2005.
12 Worker interview 2-4, Changwon, July 2005.
13 Worker interview 3-1, Changwon, July 2005.
14 Worker interview 5-2, Changwon, July 2005.
15 Worker interview 2-2, Changwon, July 2005.
16 Worker interview 3-2, Changwon, July 2005.
17 Worker interview 3-4, Changwon, July 2005.
18 Worker interview 3-5, Changwon, July 2005.
19 Worker interview 2-3, Changwon, July 2005.
20 Worker interview 1-4, Changwon, July 2005.
21 Worker interview 2-4, Changwon, July 2005.
22 Worker interview 2-1, Changwon, July 2005.
23 Worker interview 4-1, Changwon, July 2005.
24 Worker interview 2-2, Changwon, July 2005.
25 Worker interview 2-3, Changwon, July 2005.
26 Worker interview 4-3, Changwon, July 2005.
27 Worker interview 1-1, Changwon, July 2005.
28 Worker interview 1-2, Changwon, July 2005.
29 Worker interview 1-3, Changwon, July 2005.
30 Worker interview 2-3, Changwon, July 2005.
31 Worker interview 3-2, Changwon, July 2005.
32 Worker interview 3-4, Changwon, July 2005.
33 Worker interview 4-1, Changwon, July 2005.
34 Worker interview 5-1, Changwon, July 2005.
35 Interview, Shon, Seok-hyeong, Yoon, November 2005, Changwon.
36 Worker interviews 1-1, 1-2, 2-2, 2-3, 4-3, 4-4, Changwon, July 2005.
37 Worker interview 1-4, Changwon, July 2005.
38 Worker interview 3-3, Changwon, July 2005.
39 Worker interview 4-1, Changwon, July 2005.
40 Worker interview 5-2, Changwon, July 2005.

41 Worker interview 4-1, Changwon, July 2005.
42 Worker interviews 5-3 and 5-4, Changwon, July 2005.
43 Worker interviews 1-1, 1-2, 1-3 and 1-4, Changwon, July 2005.
44 Worker interviews 2-1 and 2-4, Changwon, July 2005.
45 Worker interview 3-1, Changwon, July 2005.
46 Worker interview 3-2, Changwon, July 2005.
47 Worker interview 3-4, Changwon, July 2005.
48 Worker interview 5-1, Changwon, July 2005.
49 Worker interview 5-2, Changwon, July 2005.
50 Worker interview 1-1, Changwon, July 2005.
51 Worker interviews 2-1 and 2-4, Changwon, July 2005.
52 Worker interview 4-2, Changwon, July 2005.
53 Worker interview 4-3, Changwon, July 2005.
54 Worker interview 4-1, Changwon, July 2005.
55 Worker interview 5-1, Changwon, July 2005.
56 Worker interviews 5-3 and 5-4, Changwon, July 2005.
57 Interview with Kim, Sung-Hee, and Kim, Ju-Hwan from the Korea Contingent Worker Centre, July 2005, Seoul.
58 Interview with Shin, Seng-Chul, Vice-president, KCTU, July 2005, Seoul.
59 Interview with Kim, Sung-Hee, and Kim, Ju-Hwan, July 2005, Seoul.
60 Interview with Kim, Sung-Hee, and Kim, Ju-Hwan, July 2005, Seoul.
61 Interview with Shin, Seng-Chul, vice-president, KCTU, July 2005, Seoul.
62 Interview with Shin, Seng-Chul, vice-president, KCTU, July 2005, Seoul.
63 Interview with Shin, Seng-Chul, vice-president, KCTU, July 2005, Seoul.
64 Interview with Shin, Seng-Chul, vice-president, KCTU, July 2005, Seoul.
65 Interview with Kim, Sung-Hee, and Kim, Ju-Hwan, July 2005, Seoul.
66 Worker interview 3-1, Changwon, July 2005.
67 Worker interview 3-2, Changwon, July 2005.
68 Worker interview 3-4, Changwon, July 2005.
69 Interview with Sul, Sang-seok, November 2005.
70 Interview with Shon, Seok-hyeong, November 2005.
71 Worker interview 3-4, Changwon, July 2005.
72 Worker interview 5-1, Changwon, July 2005.
73 Worker interview 3-2, Changwon, July 2005.
74 Worker interview 3-4, Changwon, July 2005.
75 Worker interview 5-1, Changwon, July 2005.

Chapter 7   Squeezing Orange

1 Lyric from the song 'Truganini' composed by Jim Moginie and Rob Hirst, from Midnight Oil's album *Earth and Sun and Moon* (2003), published by Warner Records.
2 Interview with Leon Adrewartha, the director of manufacturing for Electrolux Australia, Orange, August 2003.

3   Interview with shop floor worker, Orange, August 2003.
4   (accessed 19 October 2006).
5   Worker interview 13, Electrolux, Orange, 2005.
6   Worker interview 5, Electrolux, Orange, 2005.
7   AMWU, internal memorandum, 11 May 2004.
8   Centre Link is the Australian government agency responsible for managing the social welfare system. Unemployment benefits are currently under review, with plans to tighten them to ensure that the out of work look for work. Dennis has searched for the past eight months to no avail.
9   A member of an Australian motorcycle gang.
10 This may presage a significant political shift as these alliances emerge in other regions of Australia as family farms experience the impact of free trade agreements. The AMWU is currently supporting the protests of Tasmanian fruit and vegetable growers.
11 Extracts from the monthly newspaper *Electrolux Down Under*, Spring 2004.

Chapter 8   History Matters

1   Roughly speaking, 10 per cent of organized workers in Korea belong to industrial or regional or sectoral level unions, not enterprise-based.
2   As an observer wrote at the time: 'American influence "penetrates into every branch of administration and is fortified by an immense outpouring of money." Americans kept the government, the army, the economy, the railroads, the airports, the mines, and the factories going, supplying money, electricity, expertise, and psychological succour. American gasoline fuelled every motor vehicle in the country. American cultural influence was "exceedingly strong," ranging from scholarships to study in the United States, to several strong missionary denominations, to "a score of travelling cinemas" and theatres that played mostly American films, to the Voice of America, to big-league baseball: "America is the dream-land" to thousands if not millions of Koreans' (Cumings 1997: 255).
3   'Somewhere between 100,000 and 200,000 Korean women were mobilized into this slavery' (Cumings 1997: 179).
4   The breaches in labour standards and the attempts to publicize these breaches in the international trade union movement is movingly described by Chun Soonok (2000), younger sister of Chun Tae-il.
5   Indeed, workers' education had been central in the earlier attempts at democratic unionism with the formation of the Chonggye Labourer's Classroom in April 1975. The programme included academic subjects to primary and middle-school levels taught in accordance with the national curriculum by fully qualified volunteer teachers; vocational training courses focused upon textile and garment-making skills and aimed at facilitating the advancement of the youngest girls; a political education course consisting

of politics, economics, labour law and trade union law; a course in leadership and confidence building; and a special programme which combined students, workers and intellectuals in a 'social solidarity workshop' (Chun 2000: 305). Chun argues that the achievements of the 1970s destroyed the stereotypical image of the passive, timid, obedient and unworldly Asian female and showed them to be 'courageous, determined, compassionate and intelligent individuals capable of generating and sustaining a co-operative spirit of the highest order' (Chun 2000: 5).

6 In May 1995 it was transformed into the Korean Labour and Society Institute (KLSI) and widened to include research and policy analysis.

7 The KTUC merged with the KCIIF in 1993, an umbrella organisation for white-collar workers, to form the Korean Council of Trade Unions (KCTU). In 1995 it became the Korean Confederation of Trade Unions (KCTU).

8 Lifelong employment is a Japanese tradition, not Korean. Of course, some workers in Korea have enjoyed it, but they were in a minority – male and based in large companies. Historically, the majority of Korean workers have not enjoyed lifelong employment.

9 'The victory of [President Kim] in the presidential election in December 1997 marked the moment of democratic consolidation in South Korea' (Shin 2002: 5).

Chapter 9 Grounding Labour Internationalism

1 Adriana Rosenzvaig, address to a meeting of the AMCOR working group, Kuala Lumpur, June 10/11, 2006.

2 SIGTUR defines 'South' politically and not geographically, as those zones of the global economy that are characterized by various forms of authoritarian statism and corporate dominance. The policy document *Principles for Participation in SIGTUR* stated that only those organizations that reflected ILO Conventions 87 (freedom of association) and 98 (collective bargaining rights) are allowed to participate (SIGTUR 1999).

3 Certain Southeast Asian leaders are conscious of their family's experience as indentured labourers on plantations. They feel a sense of political betrayal as the exploitative conditions of workers remains largely unchanged while post-colonial states consolidated new elites. Part of this sense of history is shaped by the leaders' view of the United States as a dominant and politically manipulative power in the region, playing a key role in the repression of democratic unionism and the construction of pliant workplace organization.

4 We conducted two surveys of delegates at the conferences in Johannesburg in 1999 and again in Seoul in 2001. This assessment is based on the surveys as well as in-depth interviews with key leadership figures.

5 The Malaysian union leader Arokia Dass, who has played a key role in the development of SIGTUR, was detained under Malaysia's Internal Security

Act between 1987 and 1989. He was psychologically tortured, blindfolded and moved to different prisons up to three times a week. Dita Sari, a leader of the independent unions in Indonesia, was imprisoned for three years under Suharto. She campaigned against prison conditions, organizing other prisoners, and, not surprisingly, found herself placed in isolation.

6 This implies an alliance between the traditional socialist and communist movements (red) with the new social movements such as environmental activists (green).

References

Adelzadeh, A. 2003. South Africa: Human Development Report, 2003: *The Challenge of Sustainable Development: Unlocking People's Creativity*. United Nations Development Programme (UNDP). Cape Town: Oxford University Press.

African National Congress (ANC). 1994. *The Reconstruction and Development Programme*. Johannesburg: Umanyano Press.

Aguiar, L. & Herod, A. (eds). 2006. *The Dirty Work of Neoliberalism: Cleaners in the Global Economy*. Oxford: Blackwell.

Amsden, A. 1989. *Asia's Next Giant: South Korea and Late Industrialisation*. Oxford: Oxford University Press.

Appadurai, A. 2002. *The Anthropology of Politics*. Ed. J. Vincent. Oxford: Blackwell.

Australia Reconstructed: A report by the mission members to the ACTU and the TDC to Western Europe, 1987. Canberra: Government Publishing Service.

Australian Broadcasting Corporation (ABC). 2000. Background Briefing Transcript, www.abc.net.au/rn/talks/bbing/stories/s225245.htm. Accessed June 2004.

Australian Bureau of Statistics (ABS). 2004. *Economic Activities of Foreign Owned Businesses in Australia*. Canberra: Australian Government.

Australian Council of Trade Unions (ACTU). 2003. *A Fair Australia: Public Services Policy*. ACTU Congress 2003.

Australian Manufacturing Workers' Union (AMWU). 2006. Submission to the House of Representatives Employment, Workplace Relations and Workforce Participation Committee Inquiry into Employment in Automotive Component Manufacturing. Canberra: AMWU.

Auty, R. 1995. Industrial Policy Capture in Taiwan and South Korea. *Development Policy Review*, 13: 195–217.

Auty, R. (ed.). 2004. *Resource Abundance and Economic Development*. Oxford: Oxford University Press.

Bae, K.-S. and Cho, S.-J. 2004. Labor Movement. In Wonduck Lee (ed.) *Labor in Korea, 1987–2002.* Seoul: Korea Labour Institute.

Bakan, J. 2004. *The Corporation: The Pathological Pursuit of Profit and Power.* London: Constable.

Barber, B. R. 1995. *Jihad vs McWorld: How the Planet is Falling Apart and Coming Together and What This Means For Democracy.* New York: Times Books.

Barchiesi, F. 2006. *Social Citizenship and the Transformation of Wage Labour in the Making of Post-Apartheid South Africa, 1994–2001.* PhD thesis, Faculty of Humanities. Johannesburg: University of the Witwatersrand.

Bauman, Z. 1998. *Globalisation: The Human Consequences.* Cambridge: Polity Press.

Baumann, T. 1995. *An Industrial Strategy for the Electrical Durables Sector.* A Report of the Industrial Strategy Project. Cape Town: University of Cape Town Press.

Beck, U. 2000. *The Brave New World of Work.* Cambridge: Polity Press.

Beittel, M. 1992. The Witwatersrand: Black Households, White Households. In J. Smith and I. Wallerstein (eds) *Creating and Transforming Households: The Constraints of the World-Economy.* Cambridge: Cambridge University Press, pp. 197–230.

Bell, S. 1993. *Australian Manufacturing and the State: The Politics of Industry Policy in the Post-War Era.* Melbourne: Cambridge University Press.

Beynon, H. 1973. *Working for Ford.* London: Allen Lane.

Bezuidenhout, A. 2004. *Post-Colonial Workplace Regimes in the White Goods Manufacturing Industries of South Africa, Swaziland and Zimbabwe.* Doctoral thesis, Faculty of Humanities. Johannesburg: University of the Witwatersrand.

Bezuidenhout, A. 2005a. Challenging Labour Law Reforms and the Unmaking of Labour Regimes, *7th Congress of the Southern Initiative on Globalisation and Trade Union Rights (SIGTUR), Vol. 2, No. 8, Bangkok, Thailand,* 26 June–1 July. Labour Movements Research Committee (RC44), International Sociological Association.

Bezuidenhout, A. 2005b. South Africa, Swaziland and Zimbabwe – White Goods in Post-Colonial Societies: Markets, the State and Production.' In T. Nichols and S. Cam (eds) *Labour in a Global World: Case Studies from the White Goods Industry in Africa, South America, East Asia and Europe.* London: Palgrave Macmillan.

Bezuidenhout, A. 2006. What Happened to Kelvinator? The Road from Alrode to Ezakheni and Matsapha. In S. Roberts (ed.) *Sustainable Manufacturing: The Case of South Africa and Ekhurhuleni.* Cape Town: Juta.

Bezuidenhout, A. and Southall, R. 2004. International Solidarity and Labour in South Africa. In R. Munck (ed.) *Labour and Globalisation: Results and Prospects.* Liverpool: Liverpool University Press.

Bezuidenhout, A. and Webster, E. 2008. Contesting the New Politics of Space: Labour and Capital in the White Goods Industry in Southern Africa.' In

A. Herod, A. Rainnie, and S. McGrath-Champ (eds) *Handbook of Employment and Society: Working Space.* London: Edward Elgar.

Black, A. 1991. Manufacturing Development and Economic Crisis: A Reversion to Primary Production? In S. Gelb (ed.) *South Africa's Economic Crisis.* Cape Town and London: David Philip and Zed Books.

Booth, R. R. 2003. *Warring Tribes: The Story of Power Development in Australia.* Queensland: Bardak Group.

Bordogna, L. and Pedersini, R. 1999. *Collective Bargaining, Employment Security and Firm Competitiveness in Italy.* Paper presented as part of the International Labour Organisation Project on Collective Bargaining, Employment Protection/Creation and Competitiveness (available at www.msu.edu/user/block/).

Boyce, A. 1976. *The Impact of Office Reorganizations on Regional Growth Centres.* BA (Hons) thesis, Macquarie University.

Bramdaw, N. and Louw, I. 2000. ANC Alliance Has Troubled Meeting. *Business Day,* 28 August.

Brune, N., Garrett, G. and Kogut, B. 2004. The IMF and the Global Spread of Privatisation. *IMF Staff Papers,* 51 (2).

Buhlungu, S. (ed.). 2006a. *Trade Unions and Democracy: COSATU Members' Political Attitudes, 1994–2004.* Pietermaritzburg: University of KwaZulu-Natal Press.

Buhlungu, S. 2006b. Rebels Without a Cause of Their Own? The Contradictory Location of White Officials in Black Unions in South Africa, 1973–1994. *Current Sociology,* 54 (3): 427–451.

Buhlungu, S. 2006c. Whose Cause and Whose History? A Response to Maree. *Current Sociology,* 54 (3): 469–471.

Burawoy, M. 1979. *Manufacturing Consent: Changes in the Labour Process under Monopoly Capitalism.* London: Verso.

Burawoy, M. 1985. *The Politics of Production.* London: Verso.

Burawoy, M. 2001. Transition without Transformation: Russia's Involutionary Road to Capitalism. *East European Politics and Societies,* 15 (2): 269–290.

Burawoy, M. 2003. For a Sociological Marxism: The Complementary Convergence of Antonio Gramsci and Karl Polanyi. *Politics and Society,* 31 (2): 193–261.

Burawoy, M. et al. 2000. *Global Ethnography: Forces, Connections and Imaginations in a Post-Modern World.* Berkeley: University of California Press.

Castells, M. 1997. *The Power of Identity,* Vol. 2. Oxford: Blackwell.

Chang, Dae-oup. 2002. Korean Labour Relations in Transition: Authoritarian Flexibility? *Labour, Capital and Society,* 35 (1): 10–40.

Choi, J. J. 1989. *Labor and the Authoritarian State: Labor Unions in South Korean Manufacturing Industries, 1961–1980.* Seoul: Korea University Press.

Chu, Y. 2003. Labour and Democratisation in South Korea and Taiwan. *Journal of Contemporary Asia,* 28 (2): 185–202.

Chun, J. 2005. Public Dramas and the Politics of Justice: Comparison of Janitors' Union Struggle in South Korea and the United States. *Work and Occupations*, 32 (4): 486–503.

Chun, Soonok. 2000. *'They Are Not Machines': Korean Women Workers and Their Fight for Democratic Trade Unionism in the 1970s*. Comparative Labour Studies, Department of Sociology, University of Warwick.

Chun, Soon-ok. 2006. *A Single Spark: The Biography of Chun Tae-il*. Seoul: Korea Democracy Foundation.

Clark, L. 1983. Restructuring in the White Goods Industry: 1973–1983. Bachelor of Science Honours Degree thesis, School of Earth Sciences, Macquarie University.

Clegg, S., Dunphy, D. and Redding, G. (eds). 1986. *The Enterprise and Management in East Asia*. Hong Kong: University of Hong Kong Press.

Clifford, L. 2004. Edison Exits Overseas Energy. www.TheDeal.com. Accessed December 2004.

Cock, J. 2007. *The War Against Ourselves: Justice, Nature and Power*. Johannesburg: Wits University Press.

Cockett, R. 1994. *Thinking the Unthinkable: Think-Tanks and the Economic Counter-Revolution, 1931–1983*. London: Harper Collins.

Cohen, S. and Zysman, J. 1987. *Manufacturing Matters: The Myth of the Post-Industrial Economy*. New York. Basic Books.

Congress of South African Trade Unions (COSATU). 2006. *Congress Reports: Book One*. Johannesburg: COSATU.

Cooley, C. 1956. *On Self and Social Organisation*. Chicago: Chicago University Press.

Crime Information Analysis Centre (CIAC). 2005. *Crime in the RSA for the Period April to March 1994/1995 to 2003/2004*. Pretoria: South African Police Services (SAPS).

Crough, G. and Weelwright, T. 1982. *Australia: A Client State*. Victoria: Pelican Books.

Cumings, B. 1997. *Korea's Place in the Sun: A Modern History*. New York: W. W. Norton.

Daily Labour News. www.labourtoday.co.kr (only Korean).

Davies, R., Kaplan, D., Morris, M. and O'Meara, D. 1976. Class Struggle and the Periodisation of the State in South Africa. *Review of African Political Economy*, 7: 4–30.

Denoon, D. 1979. *Understanding Settler Societies*. New York: Oxford University Press.

Department of Health of South Africa. 2003. Khomanani Addresses Key Communications Issues of HIV and AIDS. Pretoria: South African Government Information.

Department of Water Affairs and Forestry (DWAF). 2003. *Strategic Framework*. Pretoria: Government Printers.

Deyo, F. 1989. *Beneath the Miracle: Labor Subordination in the New Asian Industrialism*. Berkeley: University of California Press.

Deyo, F. C. (ed.). 1987. *The Political Economy of the New Asian Industrialism.* Ithaca, NY: Cornell University Press.

Dwyer, L. 2002. Reprimand for Reebok. Extract from an email, 25 March.

Edwards, E. 1996. *Weapons to White Goods: Celebrating Email's 50 years in Orange.* Orange: Orange City Council.

EIRonline (European Industrial Relations Observatory online). 2000. On-call jobs rejected by Electrolux-Zanussi workers. www.eurofound.europa.eu/eiro/2000/07/feature/it0007159f.html. Accessed September 2007.

Electricity Supply Commission of South Africa (ESKOM). 2005. *Annual Report.* Johannesburg: ESKOM.

Electrolux Annual Reports for 1997 to 2002.

Emnambithi Local Government. 2002. *Emnambithi/Ladysmith IDP Perspective Report.* Ladysmith.

Essential Services Commission (ESC). 2003. *Commission Reports on Performance of Gas and Electricity Industries.* www.esc.vic.gov.au. Accessed June 2004.

Ewer, P., Hampson, I., Lloyd, C., Rainford, J., Rix, S., and Smith, M. 1991. *Politics and the Accord.* Sydney: Pluto Press.

Fairbrother, P. and Testi, J. 2002. The Advent of Multinational Ownership of the Victorian Electricity Generating Plants: Questions for Labour. In P. Fairbrother, M. Paddon, and J. Teicher (eds) *Privatisation, Globalisation and Labour: Studies from Australia.* Leichhardt, NSW: Federation Press.

Fairbrother, P., Paddon, M. and Teicher, J. (eds). 2002. *Privatisation, Globalisation and Labour: Studies from Australia.* Leichhardt, NSW: Federation Press.

Fakier, K. 2005. Beyond Wage Labour? The Search for Security Among Defy Workers. Honours Research Essay, Economic and Industrial Sociology Honours, School of Social Sciences, University of the Witwatersrand.

Fakier, K. 2007. The Internationalisation of the South African Labour Markets: The Need for a Comparative Research Agenda. Paper presented at a workshop on 'A Decent Work Research Agenda for South Africa'. International Institute for Labour Studies (IILS), Institute of Development and Labour Law (IDLL) and the Sociology of Work Unit (SWOP). University of Cape Town, Cape Town, 4–5 April.

Fernbach, D. 1974. *The First International and After.* London: Penguin.

Fig, D. 2005. Manufacturing Amnesia: Corporate Social Responsibility in South Africa. *International Affairs*, 81 (3): 599–617.

Fine, B. and Rustomjee, Z. 1996. *The Political Economy of South Africa: From Minerals-Energy Complex to Industrialisation.* London: Hurst.

Fine, J. 2006. *Workers Centres: Organising Communities at the Edge of the Dream.* London: ILR Press.

Foundation for Taxpayer and Consumer Rights (FTCR). 2002a. Hoax: How the Deregulation Let the Power Industry Steal $71 Billion from California. www.consumerwatchdog.org. Accessed June 2004.

Foundation for Taxpayer and Consumer Rights (FTCR). 2002b. Fact Sheet – Analysis of the Arguments for a Bailout. www.consumerwatchdog.org/utilities/fs/fs001946.php3. Accessed June 2004.

Friedman, S. 1987. *Building Tomorrow Today: African Workers in Trade Unions, 1970–1984.* Johannesburg: Ravan.

Friedman, T. 2005. *The World is Flat: A Brief History of the Globalized World in the 21st Century.* London: Allen Lane.

Fromm, E. 1947. *Man for Himself.* New York: Holt, Rinehart and Winston.

Fukuyama, F. 1991. *The End of History and the Last Man.* New York: Free Press.

Gelb, S. (ed.). 1991. *South Africa's Economic Crisis.* Cape Town and London: David Philip and Zed Books.

Gibson-Graham, J. K. 2002. Beyond Global vs. Local: Economic Politics Outside the Binary Frame. In A. Herod and M. Wright (eds) *Geographies of Power: Placing Scale.* Oxford: Blackwell.

Goldstein, A., Bonaglia, F. and Mathews, J. 2006. Accelerated Internationalization by Emerging Multinationals: The Case of White Goods. Submitted to *Journal of International Business Studies.*

Gray, K. 2006. 'Social Movement Unionism' as a Response to Neo-Liberal Restructuring: The Case of the South Korean Labour Movement. Paper presented at the International Sociological Congress, RC44. Durban, July.

Gumede, W. M. 2005. *Thabo Mbeki and the Struggle for the Soul of the ANC.* Cape Town: Struik.

Hart, G. 2002. *Disabling Globalization: Places of Power in Post-Apartheid South Africa.* Berkeley: University of California Press.

Harvey, D. 1989. *The Condition of Postmodernity: An Enquiry into the Origins of Cultural Change.* Oxford: Blackwell.

Harvey, D. 2000. *Spaces of Hope.* Edinburgh: Edinburgh University Press.

Harvey, D. 2001. *Spaces of Capital: Towards a Critical Geography.* Edinburgh: Edinburgh University Press.

Harvey, D. 2003. *The New Imperialism.* Oxford: Oxford University Press.

Harvey, D. 2006. *Spaces of Global Capitalism: Towards a Theory of Uneven Geographical Development.* London: Verso.

Harvey, E. 2003. Researching the Public's Perceptions, Views and Concerns About the Planned Installation of Pre-Paid Water Meters in Soweto. Paper presented at an African Studies Association Conference, Boston, October.

Haworth, N. and Ramsay, H. 1986. Workers of the World Untied: A Critical Analysis of the Labour Response to the Internationalization of Capital. *International Journal of the Sociology of Law and Social Policy*, 6 (2): 55–82.

Hayek, F. A. 1979. *The Political Order of a Free People.* Chicago: University of Chicago Press.

Hayward, D. 1999. 'A Financial Revolution?' The Politics of the State Budget. In B. Costar and N. Economou (eds) *The Kennett Revolution: Victorian Politics in the 1990s.* Sydney: University of New South Wales Press.

Heller, P. 1999. *The Labor of Development: Workers and the Transformation of Capitalism in Kerala, India.* Ithaca, NY: Cornell University Press.

Herod, A. 2001a. *Labor Geographies: Workers and the Landscapes of Capitalism.* New York: Guilford Press.

Herod, A. 2001b. Labour Internationalism and the Contradictions of Globalization: Or, Why the Local is Sometimes Still Important in a Global Economy. In P. Waterman and J. Wills (eds) *Place, Space and the New Labour Internationalism.* Oxford: Blackwell.

Herod, A. 2003. The Geographies of Labour Internationalism. *Social Science History,* 27 (4): 501–523.

Herod, A. and Wright, M. 2002. *Geographies of Power: Placing Scale.* Oxford: Blackwell.

Hodge, G. 2003. Privatisation: The Australian Experience. In D. Parker and D. Saal (eds) *International Handbook on Privatisation.* Williston, VT: Edward Elgar Publishing.

Hodkinson, S. 2005. Is There a Bold New Trade Union Internationalism? The ICFTU's Response to Globalization, 1996–2002. *Labour, Capital and Society,* 38 (1&2): 37–65.

Holloway, J. 2005. *Change the World Without Taking Power.* London: Pluto Press.

Hoover's. 2004. *Edison Mission Energy: Hoover's Company In-depth Records.* Website accessed December 2004.

Hunter Taskforce. 2001. Social Audit: The Impact of a Downturn in Manufacturing on People in Hunter. Research report, Hunter, Australia.

Hyman, R. 1992. *Trade Unions and the Disaggregation of the Working Class.* London: Sage.

International Labour Organization (ILO). 2004. A Fair Globalisation: The Role of the ILO. Report to the Director General on the World Commission on the Social Dimension of Globalisation. Geneva: International Labour Office.

Izaguirre, A. K. 2002. Private Infrastructure: A Review of Projects with Private Participation, 1990–2001. *Public Policy for the Private Sector,* October. World Bank: Washington, DC.

Jacka, M. and Game, A. 1980. *The White Goods Industry: The Labour Process and the Sexual Division of Labour.* School of Behavioural Sciences, Maquarie University.

Jakobsen, K. 2001. Rethinking the International Confederation of Free Trade Unions and its Inter-American Regional Organization. *Antipode,* 33 (3): 59–79.

Jenkins, C. and Klandermans, B. (eds). 1995. *The Politics of Social Protest: Comparative Perspectives on States and Social Movements.* Minneapolis: University of Minnesota Press.

Joffe, A., Maller, J. and Webster, E. 1995. South Africa's Industrialisation: The Challenge Facing Labour. In S. Frenkel and J. Harrod (eds) *Industrialisation and Labour Relations: Contemporary Research in Seven Countries*. Ithaca, NY: ILR Press.

Johnson M. and Rix S. (eds). 1993. *Water in Australia: Managing Economic, Environmental and Community Reform*. Annandale, NSW: Pluto Press.

Jose, A. V. (ed.). 2002. *Organised Labour in the 21st Century*. Geneva: International Institute of Labour Studies.

Kaplan, D. 1977. *Class Conflict, Capital Accumulation and the State: An Historical Analysis of the State in Twentieth Century South Africa*. DPhil thesis, University of Sussex.

Karan, R. 2003. Public Interest Arguments in Privatising Government Audit. *Australian Accounting Review*, 13 (2), July.

Keck, M. and Sikkink, M. 1998. *Activists beyond Borders: Advocacy Networks in International Politics*. Ithaca, NY: Cornell University Press.

Kelly, P. 1992. *The End of Certainty: The Story of the 1980s*. Sydney: Allen & Unwin.

Kenny, B. 2001. 'We are nursing these jobs': The Impact of Labour Market Flexibility on South African Retail Sector Workers. In N. Newman, J. Pape, and H. Jansen (eds) *Is There An Alternative? South African Workers Confronting Globalisation*. Cape Town: ILRIG.

KEPCO Union. www.kepco.co.kr (mainly Korean with some English information).

Kingston, M. 1999. *Off the Rails: The Pauline Hanson Trip*. St Leonards: Unwin Allen.

Korea Contingent Worker Centre. 2004. Monthly Magazine for the Temporary, Part-time, Contract and Independent Workers of Korea. *Contingent Worker*, 1 (37).

Kovel, J. 2002. *The Enemy of Nature: The End of Capitalism or the End of the World?* London: Zed Books.

KPPIU website. baljeon.nodong.net (Korean only).

Kwon, Heejung. 2004. Restructuring in Electric Industry. In Kim, Sang-gon, Kim, Gyun and Kim, Yoon-ja (eds) *Korean Electric Industry in the 21st Century*, pp. 309–373 (Korean article).

Lambert, R. 1990. Kilusang Mayo Uno and the Rise of Social Movement Unionism in the Philippines. *Labour and Industry*, 2 & 3.

Lambert, R. 1997. *State and Labour in New Order Indonesia*. Perth: University of Western Australia Press.

Lambert, R. 1998. Asian Labour Markets and International Competitiveness: Australian Transformations. *International Review of Comparative Public Policy*, 10 (special edition on Labour Markets in Transition: International Dimensions): 271–296.

Lambert, R. 1999a. Global Dance: Factory Regimes, Asian Labour Standards and Corporate Restructuring. In J. Waddington (ed.) *Globalization Patterns of Labour Resistance*. London: Mansell.

Lambert, R. 1999b. An Emerging Force? Independent Labour in Indonesia. *Labour, Capital and Society*, 32 (1): 70–107.

Lambert, R. 2005. Death of a Factory: An Ethnography of Market Rationalism's Hidden Abode in Inner-City Melbourne. *Anthropological Forum*, 14.

Lambert, R. 2006. Intellectuals and the Labour Movement. *Newsletter of the International Sociological Association Research Committee on Labour Movements*, 3 (1).

Lambert, R. & Gillan, M. 2007. "Spaces of Hope"? Fatalism, Trade Unionism, and the Uneven Geography of Capital in White Goods Manufacturing. *Economic Geography*, 83(1): 75–95.

Lambert, R. and Webster, E. 1988. The Re-emergence of Political Unionism in Contemporary South Africa. In W. Cobbet and R. Cohen (eds) *Popular Struggles in South Africa*. Trenton, NJ: Africa World Press, pp. 20–41.

Lambert, R. and Webster, E. 2001. Southern Unionism and the New Labour Internationalism. In P. Waterman and J. Wills (eds) *Place, Space and the New Labour Internationalisms*. Oxford: Blackwell.

Lambert, R. and Webster, E. 2003. Transnational Union Strategies for Civilising Globalisation. In R. Sandbrook (ed.) *Civilizing Globalization: A Survival Guide*. Albany: State University of New York Press.

Lambert, R. and Webster, E. 2004. Global Civil Society and the New Labour Internationalism: A View from the South. In R. Taylor (ed.) *Creating a Better World: Interpreting Global Civil Society*. Bloomfield: Kumarian Press.

Lambert, R. and Webster, E. 2006. Social Emancipation and the New Labour Internationalism: A Southern Perspective. In B. De Sousa Santos (ed.) *Another Production is Possible: Beyond the Capitalist Canon*. London: Verso.

Lambert, R., Gillan, M. and Fitzgerald, S. 2005. Electrolux in Australia: Deregulation, Industry Restructuring and the Dynamics of Bargaining. *Journal of Industrial Relations*, 47 (3): 261–276.

Latham, A. 2002. Re-theorizing the Scale of Globalization: Topologies, Actor Networks, and Cosmopolitanism. In A. Herod M. Wright (eds) *Geographies of Power: Placing Scale*. Oxford: Blackwell.

Lee, E. 1997. *The Labour Movement and the Internet: The New Internationalism*. London: Pluto Press.

Lee, Y. C. 1991. Kimchi: The Famous Fermented Vegetable Products in Kimchi. *Food Reviews International*, 7 (4): 399–415.

Lerner, S. 2007. Global Unions: A Solution to Labor's Worldwide Decline. *New Labor Forum*, 16 (1): 23–37.

Levinson, C. 1971. *Capitalism, Inflation and Multinationals*. London: Allen and Unwin.

Levinson, C. 1972. *International Trade Unionism*. London: Allen and Unwin.

Levinson, C. 1974. *Industry's Democratic Revolution*. London: Allen and Unwin.

Leys, C. 2001. *Market-driven Politics: Neo-liberal Democracy and the Public Interest*. London: Verso.

LG Electronics. 2005. *LG – One More Reason to Smile*. Promotional brochure.

Linge, G. 1963. The Location of Manufacturing in Australia. In A. Hunter (ed.) *The Economics of Australian Industry*. Melbourne: Melbourne University Press, pp. 18–64.

Locke, R. M. and Thelen, K. 1995. Apples and Oranges Revisited: Contextualised Comparisons and the Study of Comparative Labour Politics. *Politics and Society*, 23 (3): 337–367.

Loveday, P. 1978. Government and Manufacturing: Administrative Arrangements at the State Level. *Politics*, 13 (1): 92–103.

McNamara, M. 2006. The Hidden Health and Safety Costs of Casual Employment. Research report, Industrial Relations Centre, University of New South Wales.

Macro-Economic Research Group (MERG). 1993. *Making Democracy Work: A Framework for Macro-Economic Policy in South Africa*. Bellville: Centre for Development Studies.

Mamdani, M. 1996. *Citizen and Subject: Contemporary Africa and the Legacy of Late Colonialism*. Cape Town: David Philip.

Manne, R. 2006. The Nation Reviewed: Comment. *The Monthly*, November: 10–13.

Maree, J. 2006a. Rebels with Causes: White Officials in Black Trade Unions in South Africa, 1973–94: A Response to Sakhela Buhlungu. *Current Sociology*, 54 (3): 453–467.

Maree, J. 2006b. Similarities and Differences between Rebels With and Without a Cause. *Current Sociology*, 54 (3): 473–475.

Marquand, D. 1997. *The New Reckoning: Capitalism, States and Citizens*. Cambridge: Polity Press.

Marx, K. 1976a. *Capital*, Vol. 1. London: Penguin.

Marx, K. 1976b. *Capital*, Vol. 2. London: Penguin.

Mathews, J. 1989. *Tools of Change: New Technology and the Democratisation of Work*. Sydney: Pluto Press.

Mead, G. 1934. *Mind, Self and Society from the Standpoint of a Social Behaviourist*. Chicago: Chicago University Press.

Milkman, R. and Wong, K. 2001. Organizing Immigrant Workers: Case Studies from Southern California. In L. Turner, H. Katz, and R. Hurd (eds) *Rekindling the Movement: Labor's Quest for 21st Century Relevance*. Ithaca, NY: ILR Press.

Mills, C. Wright. 1959 [1976]. *The Sociological Imagination*. New York: Oxford University Press.

Mohammed, S. and Roberts, S. 2006. Report on the White Goods Industry to the Department of Trade and Industry. Corporate Strategy and Industrial Development Research Unit. Johannesburg: University of the Witwatersrand.

Monbiot, G. 2000. *Captive State: The Corporate Takeover of Britain*. London: Macmillan.

Monbiot, G. 2003. *The Age of Consent: A Manifesto for a New World Order.* London: Flamingo Press.

Moody, K. 1997. *Workers in a Lean World: Unions in the International Economy.* London: Verso.

Mosoetsa, S. 2003. Re-emerging Communities in Post-Apartheid South Africa: Mpumalanga Township, Kwa-Zulu-Natal, Durban. Paper presented to the Annual Workshop of the Crisis States Programme, 14–18 July. Johannesburg: University of the Witwatersrand.

Mosoetsa, S. 2005. *Micro-Level Responses to Macro-Economic Changes: Urban Livelihoods and Intra-Household Dynamics in South Africa.* PhD thesis, Faculty of Humanities. Johannesburg: University of the Witwatersrand.

Munck, R. 1988. *The New International Labour Studies: An Introduction.* London: Zed Books.

Munck, R. 2002. *Globalization and Labour: The New 'Great Transformation'.* London: Zed Books.

Munck, R. 2004. Globalisation, Labor and the Polanyi Problem, or the Issue of Counter-Hegemony. *Labor History,* 45 (3): 251–269.

Murray, C. 1981. *Families Divided: The Impact of Migrant Labour in Lesotho.* Cambridge: Cambridge University Press.

National Economic Development and Labour Council (NEDLAC). 2004. South African Trade Policy Framework: Principles and Guidelines for NEDLAC Consultations.

Nelson, M. K. and Smith, J. 1999. *Working Hard and Making Do: Surviving in Small Town America.* Berkeley: University of California Press.

Nicholls, H. 2005. *Orange Remembers Boer War 1899–1902.* Orange: Orange City Council.

Nichols, T. and Cam, S. (eds). 2005. *Labour in a Global World: Case Studies from the White Goods Industry in Africa, South America, East Asia and Europe.* Basingstoke: Palgrave Macmillan.

Nichols, T., Cam, S., Chou, W. G, Chun, S., Zhao, W. and Feng, T. 2004. Factory Regimes and the Dismantling of Established Labour in Asia: A Review of Cases from Large Manufacturing Plants in China, South Korea and Taiwan. *Work, Employment and Society,* 18 (4): 663–685.

O'Brien, R. 2000. Workers and World Order: The Tentative Transformation of the International Union Movement. *Review of International Studies,* 26: 533–555.

O'Meara, D. 1982. *Volkskapitalisme: Class, Capital and Ideology in the Development of Afrikaner Nationalism.* Johannesburg: Ravan.

OECD (Organization for Economic Cooperation and Development). 2002. Recent Privatization Trends in OECD Countries. www.oecd.org/dataoecd/29/11/1939087.pdf.

Ogden, M. 1993. *International Best Practice: A Critical Guide.* Sydney: Pluto Press.

Olle, W. and Schoeller, W. 1977. World Market Competition and Restrictions on International Trade Union Policies. *Capital and Class*, 2: 56–75.

Olle, W. and Schoeller, W. 1984. World Market Competition and Restrictions on International Trade Union Policies. In P. Waterman (ed.) *For a New Labour Internationalism*. The Hague: International Institute of Social Studies.

Peck, J. 1996. *Work-place: The Social Regulation of Labor Markets*. New York: Guilford Press.

Peetz, D. 1998. *Unions in a Contrary World: The Future of the Australian Trade Union Movement*. Cambridge: Cambridge University Press.

Peetz, D. 2006. *Brave New Workplace: How Individual Contracts are Changing Our Jobs*. Sydney: Allen & Unwin.

Pocock, B. 2006. *The Labour Market Ate My Babies*. Sydney: Federation Press.

Polanyi, K. 1967. *Dahomey and the Slave Trade: Analysis of an Archaic Economy*. Seattle: University of Washington Press.

Polanyi, K. 2001 [1944]. *The Great Transformation: The Political and Economic Origins of Our Time*. Boston: Beacon Press.

Polanyi, K., Arensburg, C. M. and Pearson, H. W. (eds). 1957. *Trade and Market in the Early Empires: Economies in History and Theory*. Glencoe, IL: Free Press.

Policy Co-ordination and Advisory Services (PCAS). 2006. *A Nation in the Making: A Discussion Document on Macro Social Trends in South Africa*. Pretoria: Presidency.

Press, P. 1989. A Critique of Trade Union Internationalism. In M. Press and D. Thomson (eds) *Solidarity for Survival: The Don Thomson Reader*. Nottingham: Spokesman.

Pusey, M. 1991. *Economic Rationalism in Canberra: A Nation Building State that Changed its Mind*. Cambridge: Cambridge University Press.

Ramsay, H. 1999. In Search of International Union Theory. In J. Waddington (ed.) *Globalization: Patterns of Labour Resistance*. London: Mansell, pp. 192–219.

Ramsay, H. and Bair, J. 1999. Working on the Chain Gang: Global Production Networks and their Implications for Organized Labour. Paper presented to the European Sociological Association, Congress on 'Will Europe Work?' 18–21 August.

Ranald, P. 1995. National Competition Policy. *Journal of Australian Political Economy*, 36, December.

Republic of South Africa. 1996. *Growth, Employment and Redistribution, a Macro-Economic Strategy*. Pretoria: Government Printers.

Rickard, J. 1984. *H. B. Higgins: A Rebel Judge*. Sydney: Allen and Unwin.

Roberts, S. (ed.) 2006. *Sustainable Manufacturing: The Case of South Africa and Ekhurhuleni*. Cape Town: Juta.

Rodrick, D. 1999. *The New Global Economy and the Developing Countries: Making Openness Work*. Washington, DC: Overseas Development Council.

Rosenthal, E. n.d. *The Durban Falkirk Story.* Durban: Unpublished manuscript.

Ross, A. (ed.). 1997. *No Sweat: Fashion, Free Trade and the Rights of Garment Workers.* London: Verso.

Rudd, K. 2006. Howard's Brutopia: The Battle of Ideas in Australian Politics. *The Monthly*, November: 46–50.

Rutherford, T. D. and Gertler, M. S. 2002. Labour in 'Lean' Times: Geography, Scale and the National Trajectories of Workplace Change. *Transactions of the Institute of British Geographers*, 27 (2): 195–212.

Sable, C., O'Rourke, D. and Fung, A. 2000. Ratcheting Labour Standards: Regulation for Continuous Improvement in the Global Workplace. Congress on Citizenship in a Global Economy. Madison: University of Wisconsin.

Sachs, W. 2002. *Fairness in a Fragile World: Memorandum for the World Summit on Sustainable Development, Johannesburg.* Berlin: Heinrich Bohl Foundation.

Sadler, D. and Fagan, B. 2004. Australian Trade Unions and the Politics of Scale: Reconstructing the Spatiality of Industrial Relations. *Economic Geography*, 80 (1): 23–43.

Sandbrook, R. (ed.). 2003. *Civilizing Globalization: A Survival Guide.* New York: State University of New York Press.

Sassoon, D. 1996. *One Hundred Years of Socialism: The West European Left in the Twentieth Century.* London: I. B. Tauris.

Segal-Horn, S., Asch, D. and Suneja, V. 1998. The Globalization of the European White Goods Industry. *European Management Journal* 16 (1): 101–109.

Seidman, G. 1994. *Manufacturing Militance: Workers' Movements in Brazil and South Africa, 1970–1985.* Berkeley: University of California Press.

Seidman, G. 2002. Deflated Citizenship: Labour Rights in a Global Order. Paper presented at the Annual Congress of the American Sociological Association, Washington.

Sen, A. 1999. *Development as Freedom.* Oxford: Oxford University Press.

Sennet, R. 1998. *Corrosion of Character: The Personal Consequences of Work in the New Capitalism.* New York: W. W. Norton.

Sharam, A. 2003. *Second Class Customers: Pre-payment Meters, the Fuel Poor and Discrimination.* North Melbourne: Energy Action Group.

Shaw, M., van Dijk, J. and Rhomberg, W. 2003. Determining Trends in Global Crime and Justice: An Overview of Results from the United Nations Surveys of Crime Trends and Operations of Criminal Justice Systems. *Forum on Crime and Society*, 3 (1 & 2): 35–63.

Sheil, C. 2002. *Water's Fall: Running the Risks with Economic Rationalism.* Sydney: Pluto Press.

Shin, K. 2002. Civilian Government and the Labor Movement in the Post-authoritarian Regime: The Case of South Korea. Paper prepared for presentation at RC44 World Congress of Sociology, July 9, Brisbane.

Shisana, O. et al. 2005. South African National HIV Prevalence, HIV Incidence, Behaviour and Communication Survey.

Silver, B. 2003. *Forces of Labour: Workers Movements and Globalization Since 1870*. Cambridge: Cambridge University Press.

Silver, B. and Arrighi, G. 2001. Workers North and South. In L. Panitch and C. Leys (eds) *Socialist Register 2001: Working Classes: Global Realities*. London: Merlin Press.

Song, H. K. 2002. Labour Unions in the Republic of Korea. In A. V. Jose (ed.) *Organised Labour in the 21st Century*. Geneva: International Institute of Labour Studies.

Sonn, H. 1997. The 'Late Blooming' of the Korean Labour Movement. *Monthly Review*, 49 (3): 117–129.

Southall, R. 1996. *Imperialism or Solidarity? International Labour and South African Trade Unions*. Cape Town: University of Cape Town Press.

Southall, R., Webster, E. and Buhlungu, S. 2006. Afterword. In S. Buhlungu (ed.) *Trade Unions and Democracy: COSATU Members' Political Attitudes, 1994–2004*. Pietermaritzburg: University of Kwa-Zulu Natal Press.

Southern Initiative on Globalization and Trade Union Rights (SIGTUR). 1998. *Founding Document*. Personal archives of Rob Lambert.

Southern Initiative on Globalization and Trade Union Rights (SIGTUR). 1999. Principles for Participation in SIGTUR. Unpublished paper. Personal archives of Rob Lambert.

Statistics South Africa. 2001. *South African Census*. Pretoria: Government Printers.

Statistics South Africa. 2004. Census 2001: Primary Tables KwaZulu-Natal: 1996 and 2001 Compared. Pretoria: Statistics South Africa. www.statssa.gov.za/publications/Report-03–02–08/Report-03–02–082001.pdf. Accessed 5 February 2007.

Swilling, M. and Russell, B. 2002. *The Size and Scope of the Non-profit Sector in South Africa*. Johannesburg: Graduate School of Public and Development Management, University of the Witwatersrand (P&DM), Centre for Civil Society (CCS).

Tarrow, S. 1994. *Power in Movement: Collective Action, Social Movements and Politics*. Cambridge: Cambridge University Press.

Tarrow, S. 2005. *The New Transnational Activism*. Cambridge: Cambridge University Press.

Tatge, M. 2000. 'How Swede It Is'. *Forbes*, 24 July: 56–57.

Teicher, J., Lambert, R. and O'Rourke, A. (eds). 2006. *Work Choices: The New Industrial Relations Agenda*. Sydney: Pearson.

Thompson, E. P. 1963. *The Making of the English Working Class*. London: Victor Gollancz.

Thomson, D. and Larson, R. 1978. *Where Were You Brother? An Account of Trade Union Imperialism*. London: War on Want.

Tilly, C. 2004. *Social Movements, 1768–2004*. Boulder: Paradigm Publishers.

Touraine, A. 1983. *Solidarity: The Analysis of a Social Movement, Poland, 1980–1981*. Cambridge: Cambridge University Press.

Turner, I. 1979. *Industrial Labour and Politics: The Dynamics of the Labour Movement in Eastern Australia 1900–1921.* Sydney: Hale & Iremonger.

Turner, L. and Hurd, R. 2001. Building Social Movement Unionism: The Transformation of the American Labour Movement. In L. Turner, H. Katz, and R. Hurd (eds) *Rekindling the Movement: Labour's Quest for Relevance in the 21st Century.* Ithaca, NY: ILR Press.

Turner, R. 1972. *The Eye of the Needle.* Johannesburg: SPRO-CAS.

Von Holdt, K. 2002. Social Movement Unionism: The South African Case. *Work, Employment and Society,* 16 (2): 283–304.

Von Holdt, K. 2003. *Transition from Below: Forging Trade Unionism and Workplace Change in South Africa.* Pietermaritzburg: University of Natal Press.

Von Holdt, K. and Webster, E. 2006. *Organising on the Periphery: New Sources of Power in the South African Workplace.* Johannesburg: Sociology of Work Unit.

Voss, K. and Sherman, R. 2000. Breaking the Iron Law of Oligarchy: Union Revitalization in the American Labour Movement. *American Journal of Sociology,* 106 (2).

Wade, R. 2003. *Governing the Market: Economic Theory and the Role of Government in East Asian Industrialization.* Princeton: Princeton University Press.

Wallerstein, I. and Smith, J. 1992. Households as an Institution of the World-Economy. In J. Smith and I. Wallerstein (eds) *Creating and Transforming Households: The Constraints of the World-Economy.* Cambridge: Cambridge University Press.

Ward, R. 1958. *The Australian Legend.* Melbourne: Oxford University Press.

Waterman, P. 1984. *For a New Labour Internationalism.* The Hague: Institute of Social Studies.

Waterman, P. 1998. *Globalization and Social Movements and the New Internationalisms.* London: Mansell.

Waterman, P. and Wills, J. (eds). 2001. *Place, Space and the New Labour Internationalisms.* Oxford: Blackwell.

Webster, E. 1985. *Cast in a Racial Mould: Labour Process and Trade Unionism in the Foundries.* Johannesburg: Ravan Press.

Webster, E. 1988. The Rise of Social Movement Unionism: The Two Faces of the Black Trade Union Movement in South Africa. In P. Frankel, N. Pines and M. Swilling (eds) *State, Resistance and Change in South Africa.* North Ryde, NY: Croom Helm, pp. 174–195.

Webster, E. and Adler, G. 1999. Towards a Class Compromise in South Africa's Double Transition. *Politics and Society,* 27 (2): 347–385.

Webster, E. and Buhlungu, S. 2004. The State of Trade Unionism in South Africa. *Review of African Political Economy,* 31 (100): 229–245.

Webster, E. and Fine, A. 1989. Transcending Traditions. In G. Moss and G. Maree (eds.) *South African Review.* Johannesburg: Ravan Press.

Webster, E. and Lambert, R. 2004. Emancipação social e o novo internaciona-lismo operário: uma perspective do Sul. In B. De Sousa Santos (ed.) *Trabalhor o mundo: Os caminhos do novo internacionalismo operário.* Porto: Edições Afrontamento.

Webster, E. and Von Holdt, K. (eds). 2005. *Beyond the Apartheid Workplace: Studies in Transition.* Pietermaritzburg: University of KwaZulu Natal Press.

Weiss, L. 1998. *The Myth of the Powerless State.* Ithaca, NY: Cornell University Press.

White, C. 1992. *Mastering Risk: Environment, Markets and Politics in Austra-lian Economic History.* Oxford: Oxford University Press.

Wilkinson, F. 1994. *Labour and Industry in the Asia-Pacific: Lessons for the Newly Industrialising Countries.* New York: Walter de Gruyter.

Williams, M. 2006. *Generative Politics: Participatory Socialist Projects in South Africa and Kerala.* Johannesburg: Department of Sociology, University of the Witwatersrand.

World Fact Book. 2004. Korea, South. www.cia.gov/cia/publications/factbook/geos/ks.html. Accessed 21 December 2005.

Wright, E. O. 2000. Working-Class Power, Capitalist-Class Interests, and Class Compromise. *American Journal of Sociology,* 105 (44): 957–1002.

Wright, E. O. 2006. A Compass for the Left. *New Left Review,* 41: 93–129.

Yoon, Hyowon. 1998. Criticism of Distorted Analysis on COSATU's Social Unionism. *Labour and Society,* 11.

Yoon, Hyowon. 2002. Strike of Three Unions, Its Progress and Responsibility. *Asia Solidarity Quarterly* (No. 9) issued by PDPD.

Yun, Aelim. 2007. The ILO Recommendation on the Employment Relationship and its Relevance in the Republic of Korea.' Global Union Research Network Working Paper. Geneva: ILO.

Interviews

Dass, Arokia, Malaysian union leader. Interviewed by Janaka Binwala, November 2001.

Fowler, Ric, International Secretary of the Construction, Mining and Energy Workers Union, Sydney, Australia. Interviewed by Rob Lambert, 23 February 2003.

Hlatswayo, Elizabeth, community worker in Ladysmith. Interviewed by Khayaat Fakier, 17 October, 2005.

Jamil, Rubina, President of the All Pakistan Federation of Trade Unions and leader of the Working Women's Organization. Interviewed by Rob Lambert, 7 November 2001.

Kim, Sung-Hee, President and Kim, Ju-Hwan, Director of Planning, Korean Contingent Workers' Centre (KCWC). Interviewed by Andries Bezuidenhout, July 2005.

Mazibuko, Duduzile, Mayor, Emnambithi Ladysmith. Interviewed by Khayaat Fakier, 18 October, 2005.

Mchunu, Cyril, local organizer of the National Union of Metal Workers of South Africa (NUMSA). Interviewed by Andries Bezuidenhout and Khayaat Fakier, 17 October 2005.

Murie, Jim, Assistant Secretary of the Communications, Electrical and Postal Union (CEPU). Interviewed by Rob Lambert, 23 February 2003.

Qwabe, Themba, Manager, Local Economic Development, Ladysmith Town Council. Interviewed by Khayaat Fakier, 21 September 2005.

Sari, Dita, leader of the independent unions in Indonesia. Interviewed by Ayu Ratih, 8 November 2001.

Shin, Seng-Chul, Vice-President, KCTU. Interviewed by Andries Bezuidenhout, July 2005.

Shon, Seok-hyeong, Chair of the Changwon branch of the DLP. Interviewed by Yoon Hyowon, November 2005.

Sul, Sang-seok, Chief Coordinator, Changwon Innovative Cluster Team. Interviewed by Yoon Hyowon, November 2005.

Index

Page numbers in *italics* indicate tables, page numbers in **bold** indicate figures.